Genomics Data Analysis

False Discovery Rates and Empirical Bayes Methods

T0250963

Genomics Data Analysis
False Discovery Rates and Empirical Bayes Methods

David R. Bickel

University of Ottawa, Ottawa Institute of Systems Biology, Department of Biochemistry, Microbiology and Immunology, Department of Mathematics and Statistics

CRC Press
Taylor & Francis Group
Boca Raton London New York

CRC Press is an imprint of the
Taylor & Francis Group, an **informa** business

A CHAPMAN & HALL BOOK

CRC Press
Taylor & Francis Group
6000 Broken Sound Parkway NW, Suite 300
Boca Raton, FL 33487-2742

First issued in paperback 2022

© 2020 by David R. Bickel
CRC Press is an imprint of Taylor & Francis Group, an Informa business

No claim to original U.S. Government works

ISBN 13: 978-1-03-247528-8 (pbk)
ISBN 13: 978-0-367-28036-9 (hbk)
ISBN 13: 978-0-429-29930-8 (ebk)

DOI: 10.1201/9780429299308

Library of Congress Cataloging-in-Publication Data

Names: Bickel, David R., author.
Title: Genomics data analysis : false discovery rates and empirical Bayes methods / David R. Bickel, University of Ottawa, Ottawa Institute of Systems Biology, Department of Biochemistry, Microbiology and Immunology, Department of Mathematics and Statistics.
Description: Boca Raton, FL : CRC Press, 2019. | Includes bibliographical references and index. | Summary: "The objective is to prepare students and scientists to analyze genomics data using empirical Bayes methods and to critically evaluate the statistical methods appearing in genomics articles. That is accomplished by providing the information needed for them to interpret p values for their current or future research"-- Provided by publisher.
Identifiers: LCCN 2019024678 (print) | LCCN 2019024679 (ebook) | ISBN 9780367280369 (hardback) | ISBN 9780429299308 (ebook)
Subjects: LCSH: Genomics--Statistical methods. | Genomics--Data processing. | Bayesian statistical decision theory.
Classification: LCC QH438.4.S73 B53 2019 (print) | LCC QH438.4.S73 (ebook) | DDC 572.8/60727--dc23
LC record available at https://lccn.loc.gov/2019024678
LC ebook record available at https://lccn.loc.gov/2019024679

Visit the Taylor & Francis Web site at
http://www.taylorandfrancis.com

and the CRC Press Web site at
http://www.crcpress.com

To my wife

An Excellent wife, who can nd?
   Her Value surpasses rare emeralds.
   Entrusted with everything by the heart of her husband, one like her will not lack beautiful plunder.
   Throughout Life, her work helps her husband.
   Yarn and wool became useful in her hands.
   The Needs of life she brings in like a ship bearing spices from overseas.
   Without fail day or night, she gave food to her household and employment to her workers.
   Recognizing a property's worth, she bought it, and from the fruit of her hands she planted a garden.
   Energized by her strength, her arms were prepared for work.
   The Night does not put out her ame, and she tasted the excellence of her labors.
   Deeds of prosperity ow from her arms, support from her hands at the spindle.
   To Anyone in need her hands are open, and she holds out fruit for the poor.
   Her Entire household she clothes, and her husband travels without worry.
   His Apparel she fashioned from linen and purple dye.
   Thought well of is he when deliberating in the chambers with the parish elders.
   Original linen aprons she made and sold to her neighbors.
   Noble and reverent are her words, spoken with discretion.
   Buoyant in the last days, she is clothed with strength and dignity.
   Impeccable is the management of her household, for she does not partake of idleness.
   The Commandments and wisdom come from her mouth, and her compassion lifted up her children and made them rich; her husband praised her.
   Many Kimmers have earned wealth, and many daughters have worked with skill, but none can hold a candle to you.
   Empty is beauty, and deceitful is attraction, for it is the lady of sense who will be commended; the fear of the Lord may she praise.
   Leave her the fruit of her hands; may her husband be praised in the chambers.

Contents

Preface

THIS BOOK IS FOR THE CLASSROOM ...

Reliable interpretation of genomics and genetic information has driven innovation in statistical methodology and its application to biological systems. Statisticians have met the need to test hundreds or thousands of hypotheses simultaneously with novel empirical Bayes methods that combine advantages of traditional Bayesian and frequentist statistics. In particular, techniques for estimating the local false discovery rate assign probabilities of differential gene expression, genetic association, etc. without requiring subjective prior distributions.

This book brings these methods to current and future scientists while keeping the mathematics at an elementary level. Readers will learn the fundamental concepts behind local false discovery rates, preparing them to analyze their own genomics data and to critically evaluate published genomics research. The book is suitable for researchers and for upper undergraduate students or beginning graduate students in the biological sciences. It gradually eases the readers into the mathematical equations needed. Exercises are provided at the end of each chapter to meet the needs of the classroom.

...AND ALSO FOR THE LAB

Researchers analyzing genomics data must use software that in most cases offers several choices of statistical methods designed for testing multiple hypotheses. Many of them are confused about whether to use methods that correct p values or whether to use various methods of false discovery rates. It is difficult not only to choose the type of method for one's own research but also to evaluate the methods chosen by other researchers.

This book will address the problem by providing current and future researchers with a solid understanding of advantages and disadvantages of various methods of multiple hypothesis testing. In particular, the book will acquaint them with benefits of empirical Bayes methods of estimating local false discovery rates. Since those methods are not always available in software, they will also learn how to convert the output of most genomics hypothesis testing software to estimates of local false discovery rates.

To keep everything as simple and practical as possible, nuance and qualifications are routinely suppressed without denying their role in certain contexts. To the best of my knowledge, this is the only book on empirical Bayes methods and local false discovery rate estimation that does not require an advanced background in mathematics and statistics.

In addition, the book offers guidance through the minefield of current criticisms of p values. While much of the book concerns converting multiple p values to estimated local false discovery rates to bypass those criticisms, it also has a chapter on the calibration of a single p value. The calibration methods explained by the book aim at a balance between ignoring problems with significance testing at one extreme and rejecting all uses of the p value at the other extreme.

OVERVIEW

Special statistical methods are needed to analyze high-dimensional data from transcriptional microarrays and from other genome-scale measurement technology. Assuming readers have some knowledge of basic biological concepts, this book gradually introduces the mathematical concepts needed to apply statistical decision and inference theory to genomics data.

The statistical methods will be explained using the same microarray examples throughout the book. Due to such repeated exposure, readers should become familiar with gene expression data, with the result that difficult statistical concepts can be understood without being obscured by unfamiliar biological concepts. The use of the same microarray data sets should also inculcate the awareness that the same data may be analyzed by more than one method.

Once the readers become familiar with the statistical methods through applications to microarray data, other data types will be covered. Related data types include genome-wide association data as well as quantitative proteomics and lipidomics data.

QUICK START FOR SELECTED TOPICS

Genetic Association

Example 1.4 has references to a website and to other parts of this book related to genetic association.

Local False Discovery Rate

Sections 3.2 and 3.4 provide introductions to local false discovery rates and their estimation. Section 5.3 states the reference class (reference set) problem. Section 8.2 explains a method of estimation called "Type II MLE" (Section 8.2.3); the prerequisites for that section are Sections 2.1 and 8.1.1.

DETAILED OUTLINE

Readers mastering the material of Chapter 1 ("Basic probability and statistics") will:

1. understand the concepts of gene expression, differential gene expression, probability mass functions, probability density functions, error models, probability distributions, normal distributions, tuples, random variables, hypothesis tests, null hypotheses, test statistics, significance levels, and p values;

2. be able to set up a contingency table from count data;

3. be able to compute a chi-square statistic on the basis of data from a contingency table;

4. understand how the above objectives are important for the analysis of gene expression data and genetic association data;

5. given genotype counts for cases and controls, be able to use the chi-square test to determine whether to reject the hypothesis of no genetic association at the 0.01 and 0.05 significance levels.

Readers mastering the material of Chapter 2 ("Introduction to likelihood") will:

1. understand the difference between a likelihood function and a probability function (§2.1);

2. be able to compute a likelihood ratio on the basis of prior probability, an error model, and data (§2.1);

3. know how to interpret a likelihood ratio as the strength of evidence for a hypothesis (§2.1);

4. understand the relationship between likelihood and the probability that a hypothesis is true (§2.3.1);

5. understand the relationship between probability and odds (§2.2);

6. be able to use likelihood to compute posterior odds on the basis of prior probability, an error model, and data (§2.3.1);

7. appreciate the importance of prior probability, especially with respect to its necessity in the interpretation of p values and likelihood ratios (§2.3.1).

Readers mastering the material of Chapter 3 ("False discovery rates") will:

1. understand the role of the likelihood ratio in defining the local false discovery rate (§3.2.2);

2. be able to estimate the nonlocal false discovery rate from a list of *p* values (§3.4);

3. understand the relationship between the LFDR and the non-local false discovery rate (§3.3);

4. understand the relationship between the LFDR and a posterior probability that a feature is affected (§3.2.1);

5. know where to find software to estimate the LFDR of each biological feature given a *p* value for each feature (§3.4).

Readers mastering the material of Chapter 4 ("Simulating and analyzing gene expression data") will:

1. understand the concept of simulating gene expression data and the relation between the posterior probability that a gene is differentially expressed and the local false discovery rate of that gene;

2. be able to estimate the posterior probability that a gene is differentially expressed given histograms of gene expression *p* values or fold change estimates;

3. be able to estimate the true mean expression level and its posterior distribution for each gene using a parametric empirical Bayes method (§4.3.3).

Readers mastering the material of Chapter 5 ("Variations in dimension and data") will:

1. know when to use a semi-parametric empirical Bayes method or a parametric empirical Bayes method in analyzing medium- to high-dimensional data;

2. be able to estimate the true mean expression level and its posterior distribution for each gene using the parametric empirical Bayes method of Section 4.3.3, even if some of the data are missing;

3. understand the concepts of a subclass and superclass and their relevance to the analysis of genomics data;

4. be able to determine whether to separate analysis of a p value subclass from analysis of a p value superclass for the purpose of estimating LFDRs (§5.3);

5. understand the biological meanings of *feature, null hypothesis*, and other terms needed to apply empirical Bayes methods of LFDR estimation to the analysis of genome-wide association data, QTL data, RT-PCR data, proteomics data, and metabolomics data.

Readers mastering the material of Chapter 6 ("Correcting bias in estimates of the false discovery rate") will:

1. understand why analyses of genomics data based on false discovery rates tend to be misleading;

2. be able to calibrate an estimated false discovery rate to correct that bias.

Readers mastering the material of Chapter 7 ("The $\mathcal{L}$ value: An estimated local false discovery rate to replace a p value") will:

1. know how to transform p values to estimates of the local false discovery rates of multiple hypotheses without relying on nonlocal false discovery rates;

2. know how to transform a single p value to an estimate of the posterior probability that the null hypothesis is true.

Readers mastering the material of Chapter 8 ("Maximum likelihood and applications") will:

1. understand the concept of maximum likelihood estimation;

2. understand the uses and limitations of p values and confidence intervals based on maximum likelihood estimation;

3. understand how maximum likelihood estimation can be used to estimate local false discovery rates;

4. understand the distinction between maximizing the likelihood function for an individual feature and maximizing the likelihood function for all features;

5. understand the analogy between maximum likelihood estimation and locating the highest point on a hill.

Finally, Appendix A extends a popular non-Bayesian method of multiple testing, and Appendix B provides concise and straightforward guidance to scientists.

FURTHER READING ETC.

A comprehensive review of empirical Bayes methods and false discovery rates would take us far beyond the scope of this book. For a more complete study of this area, I direct advanced readers to the works cited at the end of each chapter. The lists of references in the books cited give a much richer history of the domain.

Any supplementary material for this book will become available at http://davidbickel.com/genomics/ ("Statistical Inference to the Best Explanation of the Evidence"). Errata may also appear there, for I cannot control the book-wise error rate.

ACKNOWLEDGMENTS

The quotes that begin the chapters and the appendixes are from Doyle (1999). Chapter 6 develops material first released in Bickel (2016) and then in Bickel and Rahal (2019).

Now for the fun part, returning thanks to those who made the book possible. The efforts of David Grubbs, Karan Simpson, and Sherry Thomas at CRC Press made the publishing process early and efficient.

The University of Ottawa played several key roles. The book is much clearer as the result of feedback from multiple students. I thank the Department of Biochemistry, Microbiology, and Immunology and the Faculty of Medicine for granting a sabbatical to finish this project among others. The book was written in part at a writing workshop of the Centre for Academic Leadership that was organized by Françoise Moreau-Johnson.

Of course, the book is a product of my upbringing. My thrill for scientific debate comes from Bruce J. West, my PhD advisor. I deeply appreciate the sacrifices my parents made to provide for my education. The endless dice games with John and David and with my brothers Chris and Brain have left their mark.

Turning to my own family, I am grateful to my children for their encouragement to finish the book, and to the eldest for the beginning rows of Table 1.2. The flexibility and support of my wife helped me add new material just before the deadline. The poem in the dedication to her (Bickel, 2019a) is my translation of the *Prouerbia* 31:10–31 text found in Rahlfs and Hanhart (2006).

David R. Bickel
Ottawa
July 2019

CHAPTER 1

Basic Probability and Statistics

This Godfrey Norton was evidently an important factor in the matter. He was a lawyer. That sounded ominous. What was the relation between them, and what the object of his repeated visits? Was she his client, his friend, or his mistress? If the former, she had probably transferred the photograph to his keeping. If the latter, it was less likely. On the issue of this question depended whether I should continue my work at Briony Lodge, or turn my attention to the gentleman's chambers in the Temple. It was a delicate point, and it widened the field of my inquiry. I fear that I bore you with these details, but I have to let you see my little difficulties, if you are to understand the situation.

1.1 BIOLOGICAL BACKGROUND

1.1.1 Genomics Terminology

Since it is assumed that you have a knowledge of the basics of modern biology, it may be helpful to refresh your memory on (or to get up to speed with) terminology used in genomics. Knowing elementary biological concepts will help you learn the statistical concepts since the latter are explained in terms of biological examples.

1.1.2 Microarray Gene Expression

A *gene* is a unit of inherited biological information that influences various functions in an organism. A gene is said to be *expressed* if it is active in the sense of producing material that can potentially start, stop, or modify an organism's chemical, physiological, or behavioral functions. *Microarray* technology enables the simultaneous measurement of the expression level of each of thousands of genes from a biological sample drawn from a *tissue* such as the bone tissue of an animal or the stem tissue of a plant. Different tissue types express different genes under the same environmental or experimental conditions, and the same tissue types express differently under different conditions or for different individual organisms under the same conditions.

In a typical microarray experiment or observational study, the scientist interested in the effect of a drug treatment or other experimental perturbation of the biological system will measure the gene expression of two or more *treatment* individuals subjected to the perturbation and the gene expression of two or more *control* individuals not subjected to the perturbation. A gene is said to be *differentially expressed* if its true level of expression differs between hypothetical control and treatment populations of individuals from which the control and treatment groups of individuals are modeled as random samples. (A part of an organism's tissue will be called a *biological sample* to prevent confusion with a statistical *sample* as a group randomly drawn from a population that has some probability distribution.)

The *fold change* of a gene's expression is a ratio defined as follows. If a gene is more expressed in the treatment group than in the control group, the fold change is the amount of expression in the treatment group divided by the amount in the control group. Otherwise, the fold change is the amount of expression in the control group divided by the amount in the treatment group.

EXAMPLE 1.1

If the amount of a gene's expression is 4.8 times as much in the disease group as in the control group, its fold change is 4.8.

TABLE 1.1   Observed Fold Changes for Five Genes

Gene, Protein, or Metabolite Label	$\log\frac{\text{treatment}}{\text{control}}$ on Day 1	log (fold change) on Day 1	$\log\frac{\text{treatment}}{\text{control}}$ on Day 2	log (fold change) on Day 2
1	−1.9	1.9	−3.1	3.1
2	1.0	1.0	0.8	0.8
3	1.4	1.4	0.3	0.3
4	2.4	2.4	1.0	1.0
5	0.1	0.1	−1.6	1.6

Note that $\log 4.8$ could mean $\log_2 4.8 = 2.26$ or $\log_{10} 4.8 = 0.68$.

- Logarithm to base 2 $(\log_2)$ gives the number of *twofold* changes:
  - $\log_2 2 = 1; \log_2 4 = 2; \log_2 8 = 3; \log_2 1/2 = -1;$ $\log_2 1/4 = -2; \log_2 1/8 = -3; \log_2 1 = 0.$

- Logarithm to base 10 $(\log_{10})$ gives the number of *tenfold* changes:
  - $\log_{10} 10 = 1; \log_{10} 100 = 2; \log_{10} 1000 = 3;$ $\log_{10} 1/10 = -1; \log_{10} 1/100 = -2; \log_{10} 1/1000 = -3;$ $\log_{10} 1 = 0.$

Table 1.1 gives a five-gene case.

1.2 PROBABILITY DISTRIBUTIONS

Statistical analysis of microarray data and other types of genomics data is based on analogies between real measurements and random numbers. These analogies are called *sampling models* or *error models*.

A *probability distribution* is a relationship between a random number and its probability of occurring in a given interval. Here, the *probability* is approximately how many times the number lies within the interval divided by the number of its occurrences, assuming that it occurs a large number of times. Here are some important examples:

- The distribution associated with an unbiased six-sided die says the probability of rolling a 1 or 2 is 2/6 or 1/3. That means that if you roll the die 1000 times, it is likely that roughly 330 of the 1000 outcomes will be a 1 or a 2.

- Suppose that $X_{2,5}$, the base-10 logarithm of the measured gene expression (abundance level of the mRNA) in rat liver tissue in gene number 2 of rat number 5, has a *normal distribution* with a mean of 0.8 and a *standard deviation* (variability) of 1.2. In shorthand, $X_{2,5} \sim N\left(0.8, (1.2)^2\right)$, where $(1.2)^2$ is called the *variance*. The abbreviation $X_{2,5} \sim N\left(0.8, (1.2)^2\right)$ just says, "the probability that $X_{2,5}$ is 0.8 ± 1.2 is 68%, the probability that $X_{2,5}$ is $0.8 \pm (1.96)(1.2)$ is 95%, the probability that $X_{2,5}$ is $0.8 \pm (2.58)(1.2)$ is 99%, etc." (Table 1.2). Assume that all n rats treated with a certain drug have the same distribution of expression levels in gene number 2: $X_{2,1} \sim N\left(0.8, (1.2)^2\right)$, $X_{2,2} \sim N\left(0.8, (1.2)^2\right)$, $X_{2,3} \sim N\left(0.8, (1.2)^2\right)$, $X_{2,4} \sim N\left(0.8, (1.2)^2\right)$, $X_{2,5} \sim N\left(0.8, (1.2)^2\right)$, and so on. That is more briefly written as $X_{2,j} \sim N\left(0.8, (1.2)^2\right)$ for all $j = 1, 2, 3, ..., n$. Then if 1000 rats have been treated with the drug ($n = 1000$), about 680 of them have gene expression levels between -0.4 and 2.0 and about 320 of them have gene expression levels that are either smaller than -0.4 or larger than 2.0.

More generally, a random number with a normal distribution has a 68% chance of being 1 standard deviation from its mean, a 95% chance of being 1.96 standard deviations from its mean, and a 99% chance of being 2.58 standard deviations from its mean.

Section 4.4.1 shows how to generate random numbers from a normal distribution by rolling dice. That is a simple way to simulate gene expression data (§§4.2.2, 4.3).

An ordered set of numbers is called a *tuple*. For example, $\langle 1, 4, 4 \rangle$ is a 3-tuple (triple) since it is an ordered set of three numbers, and $\langle \mu, \sigma \rangle$ is a 2-tuple (double) since it is an ordered set of two numbers. Tuples are sometimes called *vectors*. A *random variable* is either a random number or a random tuple, an ordered list of random numbers.

TABLE 1.2 Abbreviations Familiar and Unfamiliar

Abbreviation	Meaning
lol	laughing out loud
rotfl	rolling on the floor laughing
brb	Be right back.
yolo	you only live once
btw	by the way
wyd	What you doing?
gm	good morning
gn	good night
$=$	is equal to
$\approx$	is approximately equal to
$X_{2,5} \sim N\left(0.8, (1.2)^2\right)$	95% chance: expression of gene 2 of rat 5 is from $0.8 - 1.96 \times 1.2$ to $0.8 + 1.96 \times 1.2$
$X_{i,j} \sim N\left(\mu_i, \sigma_i^2\right)$	95% chance: expression of gene i of rat j is from $\mu_i - 1.96\sigma_i$ to $\mu_i + 1.96\sigma_i$
$X'_{ij} \sim N\left(\mu'_i, (\sigma'_i)^2\right)$	95%: expression of gene i of *treatment* rat j is from $\mu'_i - 1.96\sigma'_i$ to $\mu'_i + 1.96\sigma'_i$
$X''_{ij} \sim N\left(\mu''_i, (\sigma''_i)^2\right)$	95%: expression of gene i of *control* rat j is from $\mu''_i - 1.96\sigma''_i$ to $\mu''_i + 1.96\sigma''_i$
$X_{i,j} \sim N\left(\mu_i, \sigma_i^2\right)$	$P_{\mu_i, \sigma_i}\left(\mu_i - 1.96\sigma_i \leq X_{i,j} \leq \mu_i + 1.96\sigma_i\right) = 95\%$
$f_2\left(3\right)$ or $P_2\left(X = 3\right)$	the probability of rolling a 3 when the bias of the die is 2
$f_\theta\left(x\right)$ or $P_\theta\left(X = x\right)$	the probability of rolling an x when the bias of the die is θ
$f_\theta\left(x\right)$ or $P_\theta\left(X = x\right)$	the probability that $X = x$ when θ is the parameter value
$f_{\mu,\sigma}\left(x\right)$	probability *density* of X at x if $\theta = \langle \mu, \sigma \rangle$ (mean μ and standard deviation σ)

EXAMPLE 1.2

This is a specific instance of a gene expression problem (Section 1.1.2). The observable level of expression of the ith gene of the jth individual subjected to the treatment is often modeled as the normally distributed random number

$$X'_{ij} \sim \mathrm{N}\left(\mu'_i, (\sigma'_i)^2\right); \quad i = 1, 2, \ldots, d; \quad j = 1, 2, \ldots, n',$$

(1.1)

where μ'_i is the unknown mean level of expression in the treatment population and where $(\sigma'_i)^2$ is the unknown variance (Table 1.2). The corresponding random number for the control individuals in the control sample is

$$X''_{ij} \sim \mathrm{N}\left(\mu''_i, (\sigma''_i)^2\right); \quad i = 1, 2, \ldots, d; \quad j = 1, 2, \ldots, n''.$$

(1.2)

Denote the abundance level of material produced by the ith gene in the jth individual of the control or treatment group by $10^{x'_{ij}}$ or $10^{x''_{ij}}$, respectively. Using angular brackets for numeric tuples and taking the base-10 logarithm of the abundance levels in the treatment group, the n'-tuple $\langle x'_{i1}, \ldots, x'_{in'} \rangle$ is modeled as a realization of the random tuple $\langle X'_{i1}, \ldots, X'_{in'} \rangle$. The n''-tuple $\langle x''_{i1}, \ldots, x''_{in''} \rangle$ and random tuple $\langle X''_{i1}, \ldots, X''_{in''} \rangle$ are defined similarly for the control group. The goal of many statistical methods of gene expression data analysis is to determine whether $\mu'_i - \mu''_i$, the level of differential expression of the ith gene, is sufficiently different from 0, for all $i = 1, 2, \ldots, d$.

1.3 PROBABILITY FUNCTIONS

The probability distribution (§1.2) of a random variable X may be represented mathematically by a *probability function* f_θ. The Greek letter θ ("theta") is the number or tuple that specifies a particular probability distribution and its probability function. That θ is called the *parameter*. For example, f_1 is the probability function for the probability distribution P_1, and f_2 is the probability function for the probability distribution P_2. That is confusing now, but it will become clear when you get to Example 1.3.

If x is discrete (integer-valued), then f_θ is a *probability mass function* (PMF), which means that $f_\theta(x)$ is simply the probability that $X = x$:

$$f_\theta(x) = P_\theta(X = x).$$

On the other hand, if x is continuous (real-valued), then f_θ is a *probability density function* (PDF). In that case, $f_\theta(x)$ is the *probability density* of X at x, which means that $f_\theta(x)$ approximates the probability that $x \leq X \leq x + \Delta$ divided by Δ for some extremely small, positive value of Δ:

$$f_\theta(x) \approx \frac{P_\theta(x \leq X \leq x + \Delta)}{\Delta},$$

where P_θ is the probability distribution specified by θ (Table 1.2). (How small is Δ? Think of any small number like 10^{-100}; Δ is smaller than that but still larger than 0.)

Indeed, the parameter θ is an index that specifies a particular probability distribution of X. For example, if $f_{\mu,\sigma}$ is the PDF that specifies $N(\mu, \sigma^2)$ as the probability distribution, then $\theta = \langle \mu, \sigma \rangle$. In other words, $f_{\mu,\sigma}(x)$ is a probability density of a normally distributed random number of mean μ and standard deviation σ.

Section 4.4 presents plots of three normal PDFs in order to clarify the concept.

EXAMPLE 1.3

Consider the hypothesis that a six-sided die is very low-biased ($\theta = -2$), slightly low-biased ($\theta = -1$), unbiased ($\theta = 0$), slightly high-biased ($\theta = 1$), or very high-biased ($\theta = 2$). Then $f_\theta(x)$ is the probability of rolling an x when the bias of the die is θ. For example, $f_2(3)$ is the probability of rolling a 3 when the bias of the die is 2 (Table 1.2). Each of those five parameter values representing five different

TABLE 1.3   The Five Columns Represent Five PMFs Corresponding to Five Hypotheses about the Bias of a Six-Sided Die

	f_{-2}	f_{-1}	f_0	f_1	f_2
f_θ (1)	50%	30%	1/6	5%	1%
f_θ (2)	30%	20%	1/6	10%	2%
f_θ (3)	10%	20%	1/6	15%	**7%**
f_θ (4)	7%	15%	1/6	20%	10%
f_θ (5)	2%	10%	1/6	20%	30%
f_θ (6)	1%	5%	1/6	30%	50%

Note: The 5-tuple of the PMFs is $\langle f_{-2}, f_{-1}, f_0, f_1, f_2 \rangle$ and is called an *error model*. Each row represents a possible outcome of a throw of the die. The numbers in the cells are probabilities; note that each column adds up to 100%, as required by the rules of probability. The **bold cell** says f_2 (3) = 7%, which means that there would be a 7% chance of rolling a 3 if the bias of the die were 2. That may also be abbreviated by $P_2 (X = 3) = 7\%$ (Table 1.2). See Example 1.3.

hypotheses about the die corresponds to a different probability distribution and thus to a different probability mass function (Table 1.3).

The PMF or PDF should never be confused with the likelihood function, which will be introduced in Section 2.1.

1.4 CONTINGENCY TABLES

Organizing information into contingency tables is an important first step for analyzing many types of Genomics data.

EXAMPLE 1.4: GENETIC ASSOCIATION STUDIES

Each somatic nucleotide of a diploid corresponds to one of two alleles denoted by "A" or "B." In a *case-control study* of a human population, a *genetic marker* records the allele inherited from each parent at a specific DNA site called a *variant* or *single-nucleotide polymorphism* (SNP). Then "AA" at a variant indicates that the A allele was inherited from both the father and the mother, "BB" indicates that the B allele was inherited from the father and mother, and "AB" indicates that the A allele was inherited from one parent and the B allele

TABLE 1.4   Contingency Table for Example 1.4: Observed Numbers of Patients with Each Genotype (AA, AB, or BB) and Disease Status (Case or Control)

	Cases	Controls	Total
AA	n_{AA}^{case}	$n_{AA}^{control}$	$n_{AA} = n_{AA}^{case} + n_{AA}^{control}$
AB	n_{AB}^{case}	$n_{AB}^{control}$	$n_{AB} = n_{AB}^{case} + n_{AB}^{control}$
BB	n_{BB}^{case}	$n_{BB}^{control}$	$n_{BB} = n_{BB}^{case} + n_{BB}^{control}$
Total	$n^{case} = n_{AA}^{case} + n_{AB}^{case} + n_{BB}^{case}$	$n^{control} = n_{AA}^{control} + n_{AB}^{control} + n_{BB}^{control}$	$n = n^{case} + n^{control} = n_{AA} + n_{AB} + n_{BB}$

from the other parent. The three *genotypes* are AA, AB, and BB. Genotypes are recorded for *cases*, patients with a certain disease, and for *controls*, patients without the disease. A *contingency table* for a case-control study has information about whether the variant is associated with the disease (Table 1.4). See http://www.genome.gov/20019523, Examples 1.5 and 5.1, and Section 1.6 for more information on large-scale genetic association studies.

A *chi-square test* is used to determine whether a contingency table indicates an association. This is accomplished by comparing *observed* counts to the counts that would be *expected* if there were no association. Let m stand for the number of cells in a contingency table, not including the totals. Also, let $\#O_1$ indicate the observed number or count in cell 1, let $\#O_2$ indicate the observed count in cell 2, and so on. Likewise, let $\#E_1$ indicate the expected count in cell 1, let $\#E_2$ indicate the expected count in cell 2, and so on. Then the *chi-square test statistic* is

$$\chi^2 = \frac{(\#O_1 - \#E_1)^2}{\#E_1} + \frac{(\#O_2 - \#E_2)^2}{\#E_2} + \cdots + \frac{(\#O_m - \#E_m)^2}{\#E_m}. \tag{1.3}$$

Higher values of χ^2 are more indicative of association and result in lower p values. Statistical software such as Excel computes the p value on the basis of the probability distribution that the chi-square statistic would have if there were no association.

In this context, a p value is the probability that the chi-square test statistic would be at least as high as χ^2 if there were no association. Since getting a very low p value would be improbable if there were no association, such a low p value would be interpreted as evidence of association. However, a high p value should not be interpreted as evidence against association.

EXAMPLE 1.5

The null hypothesis that there is no association between a variant and the disease may be assessed by the chi-square test. In this case, substituting the observed counts from Table 1.4 and the expected counts from Table 1.5 into equation (1.3) yields a chi-square test statistic of

$$
\chi^2 = \frac{\left(n_{AA}^{case} - n_{AA}n^{case}/n\right)^2}{n_{AA}n^{case}/n} + \frac{\left(n_{AA}^{control} - n_{AA}n^{control}/n\right)^2}{n_{AA}n^{control}/n}
$$

$$
+ \frac{\left(n_{AB}^{case} - n_{AB}n^{case}/n\right)^2}{n_{AB}n^{case}/n} + \frac{\left(n_{AB}^{control} - n_{AB}n^{control}/n\right)^2}{n_{AB}n^{control}/n}
$$

$$
+ \frac{\left(n_{BB}^{case} - n_{BB}n^{case}/n\right)^2}{n_{BB}n^{case}/n} + \frac{\left(n_{BB}^{control} - n_{BB}n^{control}/n\right)^2}{n_{BB}n^{control}/n}.
$$

If $\chi^2 > 6.0$, then the p value is less than 0.05, which means that the hypothesis of no association is rejected at the 0.05 significance level ($\alpha = 0.05$). If χ^2 is also greater than 9.2, then the p value is less than 0.01, which means that the same null hypothesis is rejected at the 0.01 significance level ($\alpha = 0.01$).

TABLE 1.5   Expected Numbers of Patients Derived From the Row and Column Totals of Table 1.4

	Cases	Controls
AA	$n_{AA}n^{case}/n$	$n_{AA}n^{control}/n$
AB	$n_{AB}n^{case}/n$	$n_{AB}n^{control}/n$
BB	$n_{BB}n^{case}/n$	$n_{BB}n^{control}/n$

Note: Here, n is the total number of patients: $n = n_{AA}^{case} + n_{AB}^{case} + n_{BB}^{case} + n_{AA}^{control} + n_{AB}^{control} + n_{BB}^{control}$.

1.5 HYPOTHESIS TESTS AND *P* VALUES

More generally, a *test statistic* is a number computed from data that is used to test a *null hypothesis*, which is a hypothesis that there is no effect of a drug treatment (Example 1.2), no effect of a genetic variant (Examples 1.4 and 1.5), etc. Statistical software such as Excel can convert a test statistic to a p value, which is the probability that the test statistic would be at least as high as it is if the null hypothesis were true. Except in special cases, the p value is not a probability that the null hypothesis is true. An approximation to such a probability will be presented in Chapter 3.

EXAMPLE 1.6

The null hypothesis of 0 differential gene expression would mean the treatment has no effect on the expression of the gene. The p value for a gene is a number between 0 and 1 that, based on the expression data, tends to be evenly distributed between 0 and 1 for unaffected genes and closer to 0 for affected genes. It is not necessarily the probability that the treatment has no effect on the gene expression.

If the p value is less than a pre-selected significance level such as 0.05, then the null hypothesis is *rejected*. The usual symbol for a significance level is α.

If the null hypothesis *is* rejected, there are two possible explanations of the low p value. First, the p value may be lower than the significance level because there is some evidence that the null hypothesis is false. The second explanation is very important but is usually overlooked. The p value may be lower than the significance level because α was not set low enough for the proportion of similar null hypotheses that are true. In other words, the significance level should be set at a lower level if the proportion of true null hypotheses is higher so that more probable null hypotheses are harder to reject. More precise adjustments according to that proportion will be introduced in Section 2.3.2 and Chapter 3.

TABLE 1.6   Genetic Association Data for a
Particular SNP

	Coronary Artery Disease Cases	Controls
AA	900	800
AB	500	500
BB	100	200

If the null hypothesis *is not* rejected, there are two possible explanations of the high p value. In some settings, a possible explanation is that the null hypothesis is close to the truth. However, be careful not to jump to that conclusion. A high p value often just means that not enough data have been observed to bring the p value below the significance level.

The above procedure of comparing a p value to a significance level to determine whether to reject the null hypothesis is called a *hypothesis test*.

1.6 BIBLIOGRAPHICAL NOTES

Lewis (2002), from which Examples 1.4 and 1.5 were derived, provides a useful introduction to genetic association studies.

1.7 EXERCISES (PS1–PS3)

Exercise PS1. Circle all the symbols and combinations of symbols in the book that seem confusing. Write them in a table like Table 1.2 with simple explanations that you can remember. That will help you understand the rest of this book.

Exercise PS2. Give a specific example of each of the following concepts: test statistic, null hypothesis, and significance level.

Exercise PS3. Consider the genetic association data of Table 1.6. (a) What does it mean in simple language? (b) Do you reject the hypothesis of no association between coronary artery disease and this SNP at the 0.01 significance level? Show your steps. (c) Is coronary artery disease associated with the SNP? Defend your answer.

CHAPTER 2

Introduction to Likelihood

"Indeed, your example is an unfortunate one for your argument," said Holmes, taking the paper and glancing his eye down it. "This is the Dundas separation case, and, as it happens, I was engaged in clearing up some small points in connection with it. The husband was a teetotaler, there was no other woman, and the conduct complained of was that he had drifted into the habit of winding up every meal by taking out his false teeth and hurling them at his wife, which, you will allow, is not an action likely to occur to the imagination of the average story-teller. Take a pinch of snuff, Doctor, and acknowledge that I have scored over you in your example."

2.1 LIKELIHOOD FUNCTION DEFINED

Suppose that f_θ is a probability function (PMF or PDF) as defined in Section 1.3. For some value of x fixed by observation, the *likelihood function L* is defined by

$$L(\theta) = L(\theta; x) = L_0 f_\theta(x) \qquad (2.1)$$

up to any positive constant L_0. Since the constant L_0 is not important, it may be set to a convenient value, for example, $L_0 = 1$.

The *likelihood ratio,*

$$\frac{L(\theta_1)}{L(\theta_0)} = \frac{L_0 f_{\theta_1}(x)}{L_0 f_{\theta_0}(x)} = \frac{f_{\theta_1}(x)}{f_{\theta_0}(x)},$$

measures the strength of evidence in the data x for the hypothesis that $\theta = \theta_1$ over the hypothesis that $\theta = \theta_0$. A likelihood ratio of 8 or more is considered *strong evidence,* and a likelihood ratio of 32 or more is considered *very strong evidence.* However, even if the likelihood ratio is much greater than those thresholds, the probability (density) of $\theta = \theta_1$ can easily be *lower* than the probability (density) of $\theta = \theta_0$, as will be seen in Section 2.3.

EXAMPLE 2.1

Review Example 1.3, in which the five probability functions are PMFs rather than PDFs. The error model of the die is $\langle f_{-2}, f_{-1}, f_0, f_1, f_2 \rangle$, the 5-tuple of the probability functions represented by the columns of Table 1.3. Suppose that the outcome of rolling the die is $X = 5$. Then, using $L_0 = 1$, the likelihood function is specified by the f_θ (5) row of Table 1.3: $L(-2) = 2\%$, $L(-1) = 10\%$, $L(0) = 16.7\%$, $L(1) = 20\%$, and $L(2) = 30\%$. When evaluating the evidence supporting the hypothesis that the die has a very high positive bias ($\theta = 2$) over the hypothesis that the die is unbiased ($\theta = 0$), the likelihood ratio

$$\frac{L(2)}{L(0)} = \frac{P_2(X = 5)}{P_0(X = 5)} = \frac{30\%}{16.7\%}$$

is useful as a measure of the strength of that evidence. Since the likelihood ratio is less than 8, the single roll of the die does not provide strong evidence of bias. However, there is strong evidence (but not *very* strong evidence) of a very high positive bias over a very high negative bias ($\theta = -2$):

$$\frac{L(2)}{L(-2)} = \frac{P_2(X = 5)}{P_{-2}(X = 5)} = \frac{30\%}{2\%}. \tag{2.2}$$

Nonetheless, the probability that $\theta = 2$ might be much *less* than the probability that $\theta = -2$ (Example 2.3).

2.2 ODDS AND PROBABILITY: WHAT'S THE DIFFERENCE?

The likelihood ratio is a ratio of probabilities but should not be confused with the *odds*, another ratio of probabilities. The odds that something will happen is equal to the probability that it will happen divided by the probability that it will not happen.

For instance, if the weather forecast is 20% chance of rain tomorrow, then the probability of rain is 20% (1/5), the probability of no rain is 80% (4/5), and the odds of rain is 20%. We could say that the odds of rain is 0.25 or 1-to-4, but it does not make sense to say that the odds is 25%. In the same way, the odds of no rain is 4 or 4-to-1 but not 400%.

Using elementary algebra, the probability of an event can be calculated from its odds:

$$\text{odds}(X = x) = \frac{P(X = x)}{P(X \neq x)} = \frac{P(X = x)}{1 - P(X = x)} \Rightarrow$$

$$P(X = x) = \frac{1}{1 + 1/\text{odds}(X=x)}. \tag{2.3}$$

EXAMPLE 2.2: EXAMPLES 1.3 AND 2.1, CONTINUED

The odds of rolling a 5 on a single cast of an unbiased die is

$$\text{odds}_0(X = 5) = \frac{P_0(X = 5)}{P_0(X \neq 5)} = \frac{1/6}{1 - 1/6} = \frac{1/6}{5/6} = \frac{1}{5}.$$

However, a die with a very high positive bias has an odds of rolling a 5 equal to

$$\text{odds}_2(X = 5) = \frac{P_2(X = 5)}{P_2(X \neq 5)} = \frac{30\%}{70\%} = \frac{3}{7}.$$

Table 2.1 clarifies the differences between probability, likelihood, and odds. The differences will prove crucial in Section 2.3.

TABLE 2.1   Distinguishing Marks of Probability, Likelihood, and Odds. The Die Row is Based on Examples 2.1 and 2.2

	Probability	**Likelihood Ratio**	**Odds**
rain	20%	N/A	1-to-4
die cast	$P_2 (X = 5) = 30\%$	$\frac{P_2(X=5)}{P_{-2}(X=5)} = 15$	$\frac{P_2(X=5)}{P_2(X \neq 5)} = \frac{3}{7}$
meaning	certainty	evidence	bet \$3 to win \$7
type	percentage	ratio	ratio
given	both $\longrightarrow$	random number	hypothesis
compared	nothing	hypotheses	random number

2.3 BAYESIAN USES OF LIKELIHOOD

2.3.1 Bayesian Updating

Consider the hypothesis that $\theta = \theta_1$, the hypothesis that $\theta = \theta_0$, an error model including the probability functions f_{θ_1} and f_{θ_0}, and the observation that $X = x$. Equation (3.7), which states the rule of conditional probability (Definition 3.1), implies that

$$\frac{P\left(\theta = \theta_1 | X = x\right)}{P\left(\theta = \theta_0 | X = x\right)} = \frac{P\left(\theta = \theta_1\right) \times f_{\theta_1}\left(x\right)}{P\left(\theta = \theta_0\right) \times f_{\theta_0}\left(x\right)} \tag{2.4}$$

$$= \frac{P\left(\theta = \theta_1\right)}{P\left(\theta = \theta_0\right)} \times \frac{L\left(\theta_1\right)}{L\left(\theta_0\right)}, \tag{2.5}$$

which says that

$$\left(\text{posterior odds of } \theta_1 \text{ over } \theta_0\right) = \left(\text{prior odds of } \theta_1 \text{ over } \theta_0\right)$$
$$\times \left(\text{likelihood ratio of } \theta_1 \text{ to } \theta_0\right). \tag{2.6}$$

The prior (pre-data) probabilities are $P\left(\theta = \theta_1\right)$ and $P\left(\theta = \theta_0\right)$, and the posterior (post-data) probabilities are $P\left(\theta = \theta_1 | X = x\right)$ and $P\left(\theta = \theta_0 | X = x\right)$. According to equations (2.4)–(2.6), the *prior odds* is a ratio of prior probabilities, and the *posterior odds* is a ratio of posterior probabilities. This transformation of prior odds and prior probability to posterior odds and posterior probability is called *Bayesian updating* since it brings the odds and probability up to date with the new information, the observation that $X = x$.

TABLE 2.2   Prior Distribution of the Biases of Dice. See Example 2.3

Bias Parameter	$\theta = -2$	$\theta = -1$	$\theta = 0$	$\theta = 1$	$\theta = 2$
Number of dice	100	400	999,000	400	100
Prior probability	100/ 1,000,000	400/ 1,000,000	999,000/ 1,000,000	400/ 1,000,000	100/ 1,000,000

For a general definition of *odds*, see Section 2.2. Technically, that definition only agrees with the one used in equation (2.4) if the probabilities are conditional on knowing that θ_1 and θ_0 are the only possible values of θ, for in that case $P(\theta \neq \theta_1) = P(\theta = \theta_0)$.

EXAMPLE 2.3

Continuing Examples 1.3 and 2.1, suppose that the probability distribution of dice biases follows Table 2.2. That prior distribution enables the calculation of the posterior probabilities that the die showing a certain outcome (such as $X = 5$) has a specified level of bias (such as $\theta = 2$).

According to equation (2.4), no matter how high the likelihood ratio is, the prior odds could be so small that the posterior odds is less than 1. The effect of the prior odds on the posterior odds is so strong that even the use of a rough guess at the prior odds will often lead to more reliable conclusions than attempting to interpret the likelihood ratio by itself.

2.3.2 Bounds from p Values

While the posterior odds is more informative than the likelihood ratio, a likelihood ratio is more informative than a p value. That is because the likelihood ratio can be multiplied by a prior odds to give the posterior odds, but there is no general formula for adjusting the p value according to the prior probabilities. However, treating p values as data leads to useful estimates of posterior probabilities, as will be seen in Section 3.2.

Given that those methods require multiple p values for accurate estimation, how can a single p value be interpreted? While there is no widely accepted formula for computing a likelihood ratio or posterior probability from a p value, lower bounds often hold.

To specify the bounds, consider the random number A that is 0 if the null hypothesis is true or 1 if the alternative hypothesis is true. The *alternative hypothesis* is the hypothesis that the null hypothesis is false. For example, if the null hypothesis is that the expression of the gene is unaffected by the treatment, then the alternative hypothesis is that the expression of the gene is affected by the treatment. Likewise, if the null hypothesis is that the DNA site is not associated with the disease, then the alternative hypothesis is that the DNA site is associated with the disease.

Let $p(x)$ denote the p value for testing the null hypothesis $(A = 0)$ given the observation that $X = x$, where X is random and x is observed. Let $g\left(p(x) | A = 0\right)$ represent the probability density of the p values at $p(x)$ given the null hypothesis, and let $g\left(p(x) | A = 1\right)$ represent the density given the alternative hypothesis.

Under conditions that are often reasonable in practice, the likelihood ratio has a lower bound if $p(x) \lesssim 0.37$:

$$\frac{g\left(p(x) | A = 0\right)}{g\left(p(x) | A = 1\right)} \gtrsim 2.7 \times \left|\ln p(x)\right| \times p(x).$$

In this notation, the Bayesian updating says

$$\frac{P\left(A = 0 | P(X) = p(x)\right)}{P\left(A = 1 | P(X) = p(x)\right)} = \frac{P(A = 0)\, g\left(p(x) | A = 0\right)}{P(A = 1)\, g\left(p(x) | A = 1\right)},$$

which yields this lower bound:

$$\frac{P\left(A = 0 | P(X) = p(x)\right)}{P\left(A = 1 | P(X) = p(x)\right)} \gtrsim 2.7 \times \frac{P(A = 0)}{P(A = 1)} \times \left|\ln p(x)\right| \times p(x).$$

$$(2.7)$$

If $P(A = 0) = P(A = 1) = 50\%$, then equation (2.3) gives

$$P\left(A = 0 | P(X) = p(x)\right) \gtrsim 2.7 \times \left| \ln p(x) \right| \times p(x). \qquad (2.8)$$

Whereas the prior probabilities must be known or guessed to use formula (2.7) or formula (2.8), the prior probability that a feature is affected by the treatment, disease, or other perturbation is hardly ever known in actual practice. Fortunately, that probability can be estimated, which in turn enables the estimation of the local false discovery rate (LFDR), a posterior probability that a feature is unaffected (§3.2).

In general, a method of estimating all prior probability distributions on the basis of data is called an *empirical Bayes* method. Section 4.3.3 will present such a method that is not related to the LFDR. Chapter 7 will present an LFDR method that, like the above formulas, only requires a single p value.

2.4 BIBLIOGRAPHICAL NOTES

For an introduction to the likelihood ratio as a measure of evidence and for the rationale behind using 8 and 32 as thresholds, see Royall (1997). Blume (2011) provides a more concise introduction to the likelihood approach to statistical data analysis. Following Jeffreys (1948), Bickel (2011b) suggests additional thresholds for moderate and overwhelming evidence. Applications of the likelihood paradigm to genomics data analysis and testing multiple hypotheses appear in Bickel (2012c, 2014a), Strug and Hodge (2006a,b), Strug et al. (2010), and Strug (2018). For related papers, see Section A.2.

The bounds on the likelihood ratio and posterior probability are found in Sellke et al. (2001) with exact conditions that guarantee that the bounds are respected.

The range of prior probabilities in Exercise L3 comes from Wellcome Trust Case Control Consortium (2007).

2.5 EXERCISES (L1–L3)

Exercise L1. The following questions pertain to Examples 1.3, 2.1, and 2.3. Demonstrate the correctness of your answers. (a) What is the posterior odds that the die that had an outcome of 5 has a very high positive bias as opposed to no bias? (b) What is the posterior odds that the die with an outcome of 5 has a very high positive bias as opposed to a very high negative bias?

Exercise L2. Consider the following fictitious and highly simplified scenario. Suppose that in a civil lawsuit, the *preponderance of evidence* for a hypothesis is defined as the amount of evidence needed to make that hypothesis more probable than not. Also suppose, more controversially, that "more probable than not" means "higher probability than not," where "probability" is a number such as 58%. The glove used in committing a murder fits the defendant but only fits 2% of the population. (a) Interpreting the likelihood ratio as the strength of evidence, how strong is the evidence for the hypothesis that the defendant committed the murder? *Hint:* It would be helpful to use the likelihood function with θ as the parameter, where $\theta = 1$ if the defendant committed the murder and $\theta = 0$ if not. (b) Does the preponderance of evidence support the hypothesis that the defendant committed the murder? *Hint:* Think about what percentage of the population commits murder.

Exercise L3. In a genome-wide association study (Example 5.1), between 1 in 1 million and 1 in 10,000 SNPs are associated with coronary artery disease (CAD). You find that an SNP has a likelihood ratio equal to 100 in favor of association with CAD. (a) Interpreting the likelihood ratio as the strength of evidence, how strong is the evidence for the hypothesis that the SNP is associated with CAD? (b) Is it more probable that the SNP is associated with CAD or that is not associated with CAD? (c) Now having the tools of Section 2.3.2, how would you answer Exercise PS3(c)? *Hint:* Take a peek at Section 7.1.

CHAPTER 3

False Discovery Rates

Of all the problems which have been submitted to my friend, Mr. Sherlock Holmes, for solution during the years of our intimacy, there were only two which I was the means of introducing to his notice—that of Mr. Hatherley's thumb, and that of Colonel Warburton's madness. Of these the latter may have afforded a finer field for an acute and original observer, but the other was so strange in its inception and so dramatic in its details that it may be the more worthy of being placed upon record, even if it gave my friend fewer openings for those deductive methods of reasoning by which he achieved such remarkable results. The story has, I believe, been told more than once in the newspapers, but, like all such narratives, its effect is much less striking when set forth *en bloc* in a single half-column of print than when the facts slowly evolve before your own eyes, and the mystery clears gradually away as each new discovery furnishes a step which leads on to the complete truth. At the time the circumstances made a deep impression upon me, and the lapse of two years has hardly served to weaken the effect.

3.1 INTRODUCTION

More reliable statistical methods sometimes come at the price of greater complexity of both mathematics and interpretation of the mathematics. The material of this section forms a foundation for understanding and properly using modern methods of interpreting high-dimensional data in biology.

3.2 LOCAL FALSE DISCOVERY RATE

3.2.1 Local False Discovery Rate De ned

If the fixed tuple x_i represents the observed data and the random tuple X_i represents the possible data of the ith feature out of a total of d features, then the p values are denoted by

$$p(x_1), \ldots, p(x_d),$$

where the numeric tuple x_i is an outcome of X_i. Notice that the lowercase p means p value, not probability. The uppercase P is used for probability.

EXAMPLE 3.1

Consider a microarray experiment involving the measurement of the expression of 20,000 genes in a group of two mice treated with a drug and a control group of two other mice, with pairing between treatment and control mice. If a gene is affected by the treatment, it is said to be *differentially expressed* between the treatment and control groups; otherwise, it is *equivalently expressed* in the treatment and control groups. Since each gene is a feature, the dimension is

$$d = 20{,}000.$$

If measured gene expression ratios (treatment/control) for the first gene are 2.1 for the first pair of mice and 0.9 for the second pair of mice, then $x_1 = \langle \log(2.1), \log(0.9) \rangle$ and $p(\langle \log(2.1), \log(0.9) \rangle)$ is the p value computed from $\langle 2.1, 0.9 \rangle$ as the data set to test the null hypothesis that the average ratio is 1 or, equivalently, that the mean logarithm of the expression ratio is 0. The paired or one-sample t-test in this case gives

$$p(x_1) = p\left(\langle \log(2.1), \log(0.9)\rangle\right)$$
$$= 0.59$$

as the two-sided p value. That means that the p value is 0.59 when testing the null hypothesis of equivalent expression given $\log(2.1)$ and $\log(0.9)$ as the two expression levels. See Example 5.5 for a case of much smaller d.

Let A_i stand for the random quantity that is equal to 1 if the ith feature is affected by (or associated with) the treatment, disease, or other perturbation. If the feature is unaffected (or unassociated), then A_i is equal to 0.

The *local false discovery rate* (LFDR) is the conditional probability that feature i is unaffected by the treatment, disease, or other perturbation given the p value:

$$\text{LFDR}\left(p(x_i)\right) = P\left(A_i = 0 | p(X_i) = p(x_i)\right). \qquad (3.1)$$

Thus, the LFDR is a conditional probability (Definition 3.1) of the hypothesis that feature i is unaffected given the data. Any conditional probability of a hypothesis given the data is called a *Bayesian posterior probability*. Thus, since the p value is treated as data,

$$P\left(A_i = a | p(X_i) = p(x_i)\right)$$

is a Bayesian posterior probability of the hypothesis that feature i is affected ($a = 1$) or unaffected ($a = 0$). Therefore, equation (3.1) implies that the LFDR is a Bayesian posterior probability that the feature is unaffected.

EXAMPLE 3.2

In the microarray context, a gene that is expressed differently between treatment and control groups is an affected feature, whereas a gene that is expressed equivalently between treatment and control groups is an unaffected feature. Thus,

$$P\left(A_1 = 0|X_1 = \langle \log(2.1), \log(0.9)\rangle\right)$$

is the probability that the first gene is equivalently expressed, but

$$P\left(A_1 = 1|X_1 = \langle \log(2.1), \log(0.9)\rangle\right)$$
$$= 1 - P\left(A_1 = 0|X_1 = \langle \log(2.1), \log(0.9)\rangle\right)$$

is the probability that it is differentially expressed.

3.2.2 The LFDR, Posterior Odds, and Likelihood

The posterior probability given by the LFDR of equation (3.1) delivers the posterior odds that feature i is affected by the treatment, disease, or other perturbation:

$$\frac{1 - \text{LFDR}\left(p(x_i)\right)}{\text{LFDR}\left(p(x_i)\right)} = \frac{P\left(A_i = 1|p(X_i) = p(x_i)\right)}{P\left(A_i = 0|p(X_i) = p(x_i)\right)}$$
$$= \frac{P(A_i = 1)}{P(A_i = 0)} \times \frac{g_1\left(p(x_i)\right)}{g_0\left(p(x_i)\right)}. \qquad (3.2)$$

Here, $g_1\left(p(x_i)\right)$ is the probability density of the affected features' p values at the observed p value, and $g_0\left(p(x_i)\right)$ is the probability density of the unaffected features' p values at the observed p value. The same notation will be used in Section 3.3.

For the introduction to the concept of a probability density of p values, see Section 1.3. (Figures 3.1 and 3.2 illustrate the concept of probability densities differing between affected features (black) and unaffected features (blue).) It follows that

$$\frac{g_1\left(p(x_i)\right)}{g_0\left(p(x_i)\right)}$$

is a likelihood ratio as defined in Section 2.1 when the p values are considered as data (Exercise L4). Therefore, equation (3.2) shows

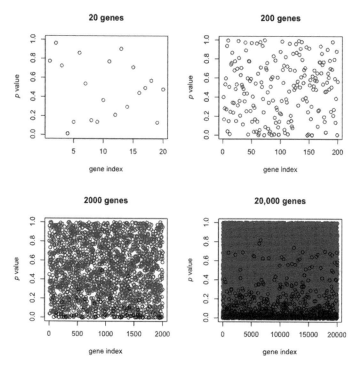

FIGURE 3.1 Simulated p values (not sorted).

that the posterior odds that a feature is affected equals the prior odds that it is affected times the likelihood ratio, as in equation (2.6). That is why the likelihood ratio is a more informative use of the p value than is comparing it to an arbitrary significance level, as in Section 1.4.

3.3 NONLOCAL AND LOCAL FALSE DISCOVERY RATES

A relationship between the LFDR and a nonlocal false discovery rate is explored in this section.

Consider a *significance level* α satisfying

$$0 < \alpha < 1$$

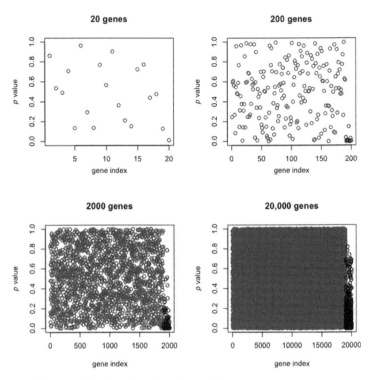

FIGURE 3.2 Simulated p values (sorted).

and the "estimate" or *prediction*

$$\hat{A}_i = \begin{cases} 1 & \text{if } p(X_i) \leq \alpha \\ 0 & \text{if } p(X_i) > \alpha \end{cases}$$

of A_i. Recall that A_i is the the random quantity equal to 1 if the ith feature is affected by a perturbation (such as a treatment or disease) and that A_i is equal to 0 otherwise. (Technically, since A_i is random, $\hat{A}_i$ is a predictor rather than an estimator of A_i.) The total number of affected features is d_1, and the total number of unaffected features is d_0; thus,

$$d_0 + d_1 = d.$$

Then the *nonlocal false discovery rate* (NFDR) is

$$\text{FDR}(\alpha) = \frac{\text{average number of false discoveries}}{\text{average number of discoveries}}$$

$$= \frac{\langle \# \left(p(X_i) \leq \alpha \text{ and } A_i = 0 \right) \rangle}{\langle \# \left(p(X_i) \leq \alpha \right) \rangle}, \tag{3.3}$$

where $\langle$angular brackets$\rangle$ denote an expectation value or mean.

EXAMPLE 3.3

In the gene expression case (Example 3.1), suppose that you choose to set $\alpha = 0.01$.

Then the NFDR is

FDR (0.01)

$$= \frac{\text{average number of false discoveries at } 0.01 \text{ significance}}{\text{average number of discoveries at } 0.01 \text{ significance}}$$

$$= \frac{\text{average number of } p \text{ values} < 0.01 \text{ for equivalently expressed genes}}{\text{average number of } p \text{ values} < 0.01}.$$

$$\tag{3.4}$$

The averages in equation (3.3) are over all possible experiments (Table 3.1). Thus, those averages cannot be known since you only have data from one experiment. For that reason, the NFDR is also unknown. However, the NFDR may be estimated by this ratio of the estimated average number of false discoveries to the number of null hypotheses rejected at significance level α:

FDR (α) estimate $= \widehat{\text{FDR}}(\alpha)$

$$= \frac{\text{estimated average number of false discoveries}}{\text{estimated average number of discoveries}}$$

$$= \frac{\text{estimated average number of false discoveries}}{\text{number of discoveries}}$$

$$= \begin{cases} \frac{\alpha d}{\#(p(x_i) \leq \alpha)} & \text{if } \frac{\alpha d}{\#(p(x_i) \leq \alpha)} < 1 \\ 1 & \text{if } \frac{\alpha d}{\#(p(x_i) \leq \alpha)} > 1. \end{cases} \tag{3.5}$$

TABLE 3.1 False-Positive Rates (FWER and NFDR) for Ten Simulated Gene Expression Experiments

	Genes in a Biochemical Pathway			All Genes		
Lab	# disc.	# false disc.	any(1)/none(0)	# disc	# false disc.	any(1)/none(0)
1	68	0	0	1208	0	0
2	100	0	0	836	37	1
3	27	1	1	1428	1	1
4	95	21	1	1876	34	1
5	83	0	0	1841	0	0
6	56	0	0	1149	23	1
7	100	7	1	1264	7	1
8	75	0	0	1405	0	0
9	99	27	1	598	484	1
10	94	0	0	735	35	1
Sum	**797**	**56**	**4**	**12340**	**621**	**7**
	False discovery rate		Family-wise e. r.	False discovery rate		Family-wise e. r.
	$\text{FDR} = 56/797$		$\text{FWER} = 4/10$	$\text{FDR} = 621/12340$		$\text{FWER} = 7/10$
	$= 7\%$		$= 40\%$	$= 5\%$		$= 70\%$

Note: FWER stands for *family-wise error rate*, a concept that leads to popular adjustments of p values such as the Bonferroni correction explained in Appendix A. The "# disc." column has the number of genes with discovered changes in expression, i.e., the number of p values less than a threshold such as 0.01 for each experiment. The "any(1)/none(0)" column has 1 if an experiment has at least one false discovery or 0 if it has no false discoveries. The **bold numbers** are used in the simplified NFDR and FWER fractions of the last row. The definitions of this table are simplified; the more precise definitions would involve an infinite number of experiments rather than only ten experiments. The relative stringency of the FWER seen in the last row is typical (Example A.1). In case the discrepancies between methods is disturbing, Appendix B is provided for guidance.

The main value of the *nonlocal* false discovery rate is to motivate the definition of its *local* counterpart.

EXAMPLE 3.4

The p value of the gene considered in Example 3.1 is $p(x_1) = 0.59$. That gene has this LFDR:

LFDR (0.59)

$$= \frac{\text{average number of false discoveries at } p \text{ value } 0.59}{\text{average number of discoveries at } p \text{ value } 0.59}$$

$$\approx \frac{\text{average number of } p \text{ values near } 0.59 \text{ forequivalently expressed genes}}{\text{average number of } p \text{ values near } 0.59}$$

LFDR (0.59)

$$= \text{probability of equivalent abundance given } p \text{ value of } 0.59.$$

Notice the similarities and differences with the NFDR as expressed in Example 3.3.

The LFDR introduced in Example 3.4 will now be defined more generally. If α is sufficiently close to 0, abbreviated as

$$\alpha \approx 0,$$

then

$$\text{FDR} (\alpha) \doteq \frac{\left\langle \# \left(p \left(X_i \right) \approx \alpha/2 \text{ and } A_i = 0 \right) \right\rangle}{\left\langle \# \left(p \left(X_i \right) \approx \alpha/2 \right) \right\rangle}$$

$$= \frac{P \left(p \left(X_i \right) \approx \alpha/2 \text{ and } A_i = 0 \right)}{P \left(p \left(X_i \right) \approx \alpha/2 \right)} \quad (3.6)$$

for all $i = 1, \ldots, d$. Here, the "$\doteq$" and "$\approx$" mean *is approximately equal to*.

Definition 3.1

The vertical bar ("|") means *if, given that*, or *conditional on*. The *conditional probability* P (event 1|event 2) is pronounced "the probability of event 1 given event 2" and means *the probability of event 1 if event 2 happens*. Conditional probability satisfies the rule

$$P \text{ (event 1|event 2)} = \frac{P \text{ (event 1 and event 2)}}{P \text{ (event 2)}}, \quad (3.7)$$

which may also be written as

$$P \text{ (event 1 and event 2)} = P \text{ (event 1|event 2)} P \text{ (event 2)}.$$

Fact 3.1

According to the law of total probability,

$$P\,(\text{event 3}) = P\,(\text{event 1})\,P\,(\text{event 3}|\text{event 1})$$
$$+ P\,(\text{event 2})\,P\,(\text{event 3}|\text{event 2}).$$

By applying Definition 3.1 to equation (3.6), you can see for yourself that

$$\text{FDR}\,(\alpha) \doteq \frac{P\,(A_i = 0)\,P\left(p\,(X_i) \approx \alpha/2 | A_i = 0\right)}{P\left(p\,(X_i) \approx \alpha/2\right)}. \tag{3.8}$$

Define probability densities $g_0\,(\alpha/2)$, $g_1\,(\alpha/2)$, and $g\,(\alpha/2)$ to satisfy

$$g\,(\alpha/2) \doteq \frac{P\left(p\,(X_i) \approx \alpha/2\right)}{\alpha},$$

$$g_0\,(\alpha/2) \doteq \frac{P\left(p\,(X_i) \approx \alpha/2 | A_i = 0\right)}{\alpha},$$

$$g_1\,(\alpha/2) \doteq \frac{P\left(p\,(X_i) \approx \alpha/2 | A_i = 1\right)}{\alpha}.$$

The law of total probability (Fact 3.1) then leads to

$$\frac{d_0}{d}g_0\,(\alpha/2) + \frac{d_1}{d}g_1\,(\alpha/2) = g\,(\alpha/2)$$
$$P\,(A_i = 0)\,g_0\,(\alpha/2) + P\,(A_i = 1)\,g_1\,(\alpha/2) = g\,(\alpha/2).$$

Then, using equation (3.8),

$$\text{FDR}\,(\alpha) \doteq \frac{P\,(A_i = 0)\,P\left(p\,(X_i) \approx \alpha/2 | A_i = 0\right)/\alpha}{P\left(p\,(X_i) \approx \alpha/2\right)/\alpha}$$

$$\doteq \left(\frac{d_0}{d}\right)\frac{g_0\,(\alpha/2)}{g\,(\alpha/2)}.$$

Accordingly, for any p between 0 and 1,

$$\text{LFDR}\,(p) = \left(\frac{d_0}{d}\right)\frac{g_0\,(p)}{g\,(p)} \tag{3.9}$$

is an LFDR. Also, the law of total probability (Fact 3.1) now gives

$$\frac{d_0}{d} g_0 \left(p\right) + \frac{d_1}{d} g_1 \left(p\right) = g \left(p\right)$$

$$P \left(A_i = 0\right) g_0 \left(p\right) + P \left(A_i = 1\right) g_1 \left(p\right) = g \left(p\right).$$

More simply, the NFDR evaluated at significance level α is approximately equal to the average LFDR over all genes with p values less than α if d is large enough. In other words,

$$\text{FDR} \left(\alpha\right) \approx \frac{\text{LFDR} \left(p_1\right) + \text{LFDR} \left(p_2\right) + \cdots + \text{LFDR} \left(p_{\#\left(p(x_i)\leq\alpha\right)}\right)}{\# \left(p_i \leq \alpha\right)},$$

$$(3.10)$$

in which $p_1, p_2, \ldots, p_{\#\left(p_i \leq \alpha\right)}$ are all less than α.

3.4 COMPUTING THE LFDR ESTIMATE

Many estimators of the LFDR have been encoded in free software, most notably in the R statistical computing language (R Development Core Team, 2008). For example, there are LFDR estimators in the R packages `locfdr`, `fdrtool`, `LFDR.MLE`, and `PsiHat` available from the Comprehensive R Archive Network (CRAN) at http://cran.r-project.org. Links to more recent R code for LFDR estimation may be found at http://bit.ly/2kcGh3O (or http://davidbickel.com → "software" category).

Relatively simple ways to compute estimates of the LFDR will be explained in Chapter 6 and Section 8.2.3. Guidelines for choosing between methods appear in Appendix B.

3.5 BIBLIOGRAPHICAL NOTES

Benjamini and Hochberg (1995) provided an algorithm that controls the original *false discovery rate* (FDR). Efron et al. (2001) applied the FDR to the analysis of gene expression microarray data and introduced the LFDR. Genovese and Wasserman (2003) and

Efron and Tibshirani (2002) pointed out important connections between the FDR, the LFDR, and Bayesian statistics.

Terminology from Bickel (2013) was adopted here with minor changes for the sake of clarity. Efron et al. (2001) based the name of the LFDR on a relationship to the false discovery rate of Benjamini and Hochberg (1995), but it is more closely related to the NFDR. Efron and Tibshirani (2002) called the NFDR the *Bayesian FDR*.

The `fdrtool` R package is described in Strimmer (2008).

3.6 EXERCISES (L4; A–B)

Exercise L4. Section 3.2.2 claims that $g_1\left(p\left(x_i\right)\right)/g_0\left(p\left(x_i\right)\right)$ is a likelihood ratio. Demonstrate that by filling in the missing steps. *Hint:* The demonstration will include the use of a likelihood function with a_i as the parameter, where $a_i = 1$ if feature i is affected and $a_i = 0$ if feature i is unaffected.

Exercise A. Suppose that A is the random quantity of a coin flip such that its value a is either heads ($a = 0$) or tails ($a = 1$) and that the value of a is not observed. The coin is fair in the sense that heads and tails occur with equal frequency. You want to compute the probability that the coin landed heads based on the observed outcome of the roll of a six-sided die. Use x to abbreviate the number of the observed outcome; x could be 1, 2, 3, 4, 5, or 6. If the coin landed heads, a fair six-sided die was rolled, but if the coin landed tails, a loaded six-sided die was rolled. Thus, the observed number x is an outcome of X, a random number of probability distribution P_a, where

$$P_0\left(X = x\right) = \frac{1}{6}$$

and

$$P_1\left(X = x\right) = \begin{cases} \frac{1}{8} & \text{if } x = 1, 2, 3, \text{ or } 4 \\ \frac{1}{4} & \text{if } x = 5 \text{ or } 6 \end{cases}$$

for $x = 1, 2, 3, 4, 5, 6$.

a. Suppose the outcome of rolling the die is 2. What is the Bayesian posterior probability that the coin landed heads? Show the steps needed to demonstrate the correctness of your answer.

b. Now suppose instead that the outcome of rolling the die is 5. What is the Bayesian posterior probability that the coin landed heads? Show the steps needed to demonstrate the correctness of your answer.

c. Use this coin-dice error model with d coin flips and d corresponding observations of corresponding die outcomes as an analogy in order to explain the application of the LFDR to a gene expression data set like that of Examples 3.1 and 3.2:

   i. What biological concept does the coin flip represent?

   ii. What biological concept do the dice outcomes represent?

   iii. What biological concept do the Bayesian posterior probabilities represent?

Exercise B. Assume that none of the 20,000 features is affected by the perturbation, i.e., all 20,000 features are unaffected. What is the value of the LFDR for each of the features? Demonstrate the correctness of your answer by showing your steps.

CHAPTER 4

Simulating and Analyzing Gene Expression Data

I trust that I am not more dense than my neighbours, but I was always oppressed with a sense of my own stupidity in my dealings with Sherlock Holmes. Here I had heard what he had heard, I had seen what he had seen, and yet from his words it was evident that he saw clearly not only what had happened but what was about to happen, while to me the whole business was still confused and grotesque. As I drove home to my house in Kensington I thought over it all, from the extraordinary story of the red-headed copier of the *Encyclopædia* down to the visit to Saxe-Coburg Square, and the ominous words with which he had parted from me. What was this nocturnal expedition, and why should I go armed? Where were we going, and what were we to do? I had the hint from Holmes that this smooth-faced pawnbroker's assistant was a formidable man—a man who might play a deep game. I tried to puzzle it out, but gave it up in despair and set the matter aside until night should bring an explanation.

4.1 SIMULATING GENE EXPRESSION WITH DICE

Statisticians simulate random numbers representing gene expression to test statistical methods of identifying affected genes. A good simulation is mathematically equivalent to a biologically realistic game. Such a game generates artificial data similarly to how experiments or observational studies generate biological data.

For speed and flexibility, statisticians normally perform their simulation studies by writing computer programs rather than flipping coins or spinning roulette wheels. However, to shed light on the type of statistical reasoning seen in Chapter 3, simulation algorithms will be described in terms of simple dice games in Sections 4.2 and 4.3. Section 4.4 explains normal distributions to bridge the two approaches while clarifying the idea of a probability density.

These standard dice abbreviations will prove useful:

- 1d4 = roll one four-sided d̲ie;

- 10d20 = roll ten twenty-sided d̲ice, and add the results;

- 5d20L3 = roll five twenty-sided d̲ice, and add the results of the lowest three dice;

- 2d20L1 = roll two twenty-sided d̲ice, and take the lowest one.

The dice are considered unbiased in the sense that each side or face of a die has equal probability of occurring. The dice games are played between two competing teams:

1. *Nature* generates numbers but hides some of them from the other team.

   a. The hidden numbers are *parameter values*.

   b. The other numbers are *observed data*.

2. *Lab* estimates the parameter values based on the observed data.

Every gene game is defined by the rules followed by each team and by the criterion for winning.

Let n denote the number of animals, people, or other individuals in the disease group, and assume that the control group has the same number of individuals. Every disease individual is paired with a different control individual, resulting in n disease/control expression ratios. Each ratio is the amount of observed expression in the disease group divided by the amount of observed expression in the control group, as in Table 4.1, which uses $n = 2$ to represent an individual from each group measured on two different days and which uses "treatment" rather than "disease" as the perturbation. In each DE game (defined in Section 4.2), Nature tells Lab the *observed expression level* $\bar{x}_i$, which is the averaged logarithm of the expression ratio for each gene:

$$
\begin{aligned}
\bar{x}_i &= \text{averaged } \log_2 (\text{disease}/\text{control}) \\
&= \frac{x_{i1} + \cdots + x_{in}}{n},
\end{aligned}
\tag{4.1}
$$

where, for the ith gene, x_{i1} is the logarithm of the disease-to-control expression ratio in disease-control pair 1, x_{i2} is the logarithm of the disease-to-control expression ratio in pair 2, ..., and x_{i1} is the logarithm of the disease-to-control expression ratio in pair n, the last disease-control pair. In statistical terminology, each $\bar{x}_i$ is a *sample mean* since it is equal to the mean of a sample of observations. Equation (4.1) says

$$
\bar{x}_1 = \frac{x_{1,1} + \cdots + x_{1,n}}{n}, \ldots, \bar{x}_d = \frac{x_{d,1} + \cdots + x_{d,n}}{n}.
$$

That will make more sense after plugging in some numbers. Table 4.1 displays possible data for the experimental design of Example 3.1 but with only five genes measured instead of the 20,000 genes of the microarray.

TABLE 4.1   Gene Expression Data for Five Genes ($d = 5$)

Gene	x_{i1}	x_{i2}	$\bar{x}_i$
$i = 1$	$x_{1,1} = -1.9$	$x_{1,2} = -3.1$	$\bar{x}_1 = -2.5$
$i = 2$	$x_{\_,\_} = 1.0$	$x_{\_,\_} = 0.8$	$\bar{x}_\_ = 0.9$
$i = 3$	$x_{\_,\_} = 1.4$	$x_{\_,\_} = 0.3$	$\bar{x}_\_ = 0.8$
$i = 4$	$x_{\_,\_} = 2.4$	$x_{\_,\_} = 1.0$	$\bar{x}_\_ = 1.7$
$i = 5$	$x_{\_,\_} = 0.1$	$x_{\_,\_} = -1.6$	$\bar{x}_\_ = -0.8$

Note: The last column is the sample mean over the observed logarithms of expression ratios. See Example 5.5. The successful student will fill in the blanks (Exercise D).

4.2 DE GAMES

Here, a disease is used instead of a treatment as the perturbation. A gene is differentially expressed (DE) if its expression is affected by the disease but is equivalently expressed if its expression is unaffected by the disease. Team Lab wants to analyze measurements of gene expression to determine which genes are DE, that is, affected by the disease.

For the games of this section, consider the random number A_i that is 0 if the expression of the ith of d genes is unaffected by a disease or 1 if it is affected by the disease. (As in Chapter 3, d stands for the *dimension* of the data.) For any $i = 1, \ldots, d$, the probability that gene i is affected is the same value, which will be abbreviated as $P(1) = P(A_i = 1)$.

There are three *settings* for the DE games:

1. Unrealistic simulation: Lab knows that $P(1) = 10\%$.

2. More realistic simulation: Nature chooses $P(1)$ to be 0%, 5%, 10%, 15%, etc., not telling Lab the value.

3. Most realistic simulation: Nature chooses $P(1)$ randomly with $P(1) = 0\%$ possible (that would mean that all genes are equivalently expressed), not telling Lab the value.

In short, Nature chooses a value of $P(1)$ to be 0%, 5%, 10%, 15%, ..., or 100% and tells it to Lab in Setting 1 but not in Setting 2 or

TABLE 4.2   Scoring System for the DE Games

	Each Discovery	Each Non-discovery
Affected genes	True discovery (+1 for Lab)	False negative
Unaffected genes	False discovery (+10 for Nature)	True negative

Setting 3. The information Nature tells Lab also varies between the DE game defined in Section 4.2.1 and the one defined in Section 4.2.2.

With that information, Lab then chooses which genes to "discover" as affected $(\widehat{A}_i = 1)$ and which to not discover $(\widehat{A}_i = 0)$. Lab may use the concepts of previous sections to make its decisions, as illustrated in the exercises in Section 4.6.

The winner is determined by the scoring system of Table 4.2. This means that Lab should only claim a discovery when the posterior odds of differential expression are at least 10-to-1.

4.2.1 Basic DE

In the game called *Basic DE*, Nature follows these steps for every gene i from 1 to d:

1. A twenty-sided die (1d20) determines whether the gene is affected $(A_i = 1)$ or unaffected $(A_i = 0)$ according to Table 4.3. This A_i value is not told to Lab.

2. A four-sided die determines whether disease *seems* to decrease (1d4 = 1,2) or increase (1d4 = 3,4) the expression of the gene. Nature tells Lab this value.

3. An eight-sided die (1d8) determines how many *observed* twofold changes in expression occur from the control group to the disease group, averaged over n ratios in each group. Nature tells Lab this fold change. The gene has less observed

TABLE 4.3   Is the *i*th Gene Affected by the Disease?

$P(1)$	5%	10%	15%	⋯	100%
Needed for $A_i = 1$	1d20 = 20	1d20 ≥ 19	1d20 ≥ 18	⋯	1d20 ≥ 1

expression in the disease group than in the control group if the outcome of rolling a four-sided die is 1 or 2 but more expression if the outcome is 3 or 4. Thus, the logarithm of the averaged expression ratio is

$$\bar{x}_i = \begin{cases} -1d8 & \text{if } 1d4 = 1 \text{ or } 2 \\ +1d8 & \text{if } 1d4 = 3 \text{ or } 4. \end{cases}$$

If the gene appears to be more expressed in the control group, the expression ratio will be less than 1 on average, making $\bar{x}_i < 0$.

4. Nature tells Lab the value of $p(x_i)$, the p value calculated this way:

   a. If the gene is unaffected (equivalently expressed), then its p value is determined by rolling two ten-sided dice, one for each digit (1d100%).

   b. If the gene is affected (differentially expressed), then its p value is

      i. 2d100L1% given two individuals per disease/control group (i.e., if $n = 2$),

      ii. 3d100L1% given three individuals per disease/control group (i.e., if $n = 3$),

      iii. 4d100L1% given four individuals per disease/control group (i.e., if $n = 4$),

      iv. nd100L1% given n individuals per disease/control group.

4.2.2 Advanced DE

In the game called *Advanced DE*, Nature follows these steps for every gene i from 1 to d:

1. This step is the same as Step 1 of Basic DE. See Section 4.2.1.

2. Nature chooses the value of a *scale* (or "spread") parameter s_i for the *i*th gene. The scale s_i quantifies the variability of $\bar{x}_i$, the logarithm of the expression ratio averaged over the *n* expression ratios. Each of the *d* genes may have a different scale, some number between 0.1 and 1. For example, if there are ten genes ($d = 10$), the ten scales could be $s_1 = 1, s_2 = 3, \ldots, s_{10} = 0.7$. Nature tells Lab those values.

3. Nature determines the *unexplained variation* y_i, which is the change in gene expression due to factors other than the presence or absence of disease. This is done by adding up the outcomes of three dice of six sides each (3d6) and applying this formula:

$$y_i = (3\text{d}6 - 10) \times \frac{s_i}{\sqrt{n}}.$$

Nature does not tell Lab that value.

4. Nature uses the value of A_i from Step 1 and the value of y_i from Step 2 to determine $\bar{x}_i$ by rolling 1 die of 12 sides (1d12) and applying this formula:

$$\bar{x}_i = \begin{cases} \bar{x}_i = y_i & \text{if } A_i = 0 \\ \bar{x}_i = y_i + 3 \times s_i & \text{if } A_i = 1 \text{ and } 1\text{d}12 \geq 7 \\ \bar{x}_i = y_i - 3 \times s_i & \text{if } A_i = 1 \text{ and } 1\text{d}12 \leq 6. \end{cases}$$

Explanation: Whereas $A_i = 0$ means that the disease does not change the expression level of the gene, $A_i = 1$ means that the disease either increases (1d12 $\geq$ 7) or decreases (1d12 $\leq$ 6) the expression level by an amount equal to 3 $\times$ s_i. Nature tells Lab the value of $\bar{x}_i$.

With the values supplied by Nature, Lab should rescale the averaged expression levels in order to decide which genes to "discover" as affected $(\widehat{A}_i = 1)$ and which to not discover $(\widehat{A}_i = 0)$. The rescaled average expression levels are given by

$$z_i = \text{rescaled } \overline{x}_i = \frac{\overline{x}_i}{3s_i/\sqrt{n}} = \frac{\overline{x}_i\sqrt{n}}{3s_i} \tag{4.2}$$

for every gene i from 1 to d.

A conventional method of determining $\widehat{A}_i$ is to compare each z_i with the z_i values that correspond to p values equal to $\alpha = 1\%$ or $\alpha = 5\%$:

$$p(x_i) \approx \begin{cases} 32\% & \text{if } z_i = \pm 1 \\ 5\% & \text{if } z_i = \pm 1.96 \\ 1\% & \text{if } z_i = \pm 2.58. \end{cases}$$

Thus, following the terminology of Section 1.4, the null hypothesis of no disease effect $(A_i = 0)$ is rejected $(\widehat{A}_i = 1)$ at the 5% significance level if $|z_i| > 1.96$ and at the 1% significance level if $|z_i| > 2.58$. If the no-effect hypothesis is not rejected at level α, then $\widehat{A}_i = 0$.

A more reliable method of determining $\widehat{A}_i$ is illustrated in the exercises in Section 4.6.

4.3 EFFECTS AND ESTIMATES (E&E)

4.3.1 Contrast with the DE Games

In the DE games (§4.2), Lab's goal is to determine which genes are differentially expressed (affected by the disease). By contrast, in the game called *Effects and Estimates (E&E)*, the goal of Lab is instead to estimate the size of the disease effect, that is, the true mean expression level of each gene.

4.3.2 Generating Mean Expression Levels

Whereas $\overline{x}_i$ is the *observed* mean expression level in the ith gene, μ_i is the *true* mean expression in the ith gene. For very large values of n, the number of individuals, $\overline{x}_i$ tends to be close to μ_i. The difference between $\overline{x}_i$ and μ_i will become clear by playing E&E.

In the game, Nature first chooses two parameter values related to the scale or spread:

1. Nature chooses the value of a *within-gene standard deviation* parameter σ for the variability of $\bar{x}_i$ within each gene. Each of the d genes has the same σ, a single number between 0.1 and 4. Nature tells Lab σ, the value of the within-gene standard deviation.

2. Nature chooses the value of a *between-gene standard deviation* parameter S for the variability of μ_i between genes. Each of the d genes has the same S, a single number between σ and 4. Nature does not tell Lab S, the value of the between-gene standard deviation.

Nature then follows these steps for every gene i from 1 to d:

1. Nature chooses the value of the *overall average expression level m*, which is the mean change in gene expression level averaged over all of the genes. (In many real experiments, $m = 0$ is a good approximation, so it is a good choice the first time the game is played.) Nature does not tell Lab the value chosen for m.

2. Nature determines the *effect size* μ_i, which is the mean change in the gene expression level that is an effect of the disease, that is, the change can be explained by the disease. This is done by adding up the outcomes of three dice of six sides each (3d6) and applying this formula:

$$\mu_i = m + (3d6 - 10) \times \frac{S}{3}. \qquad (4.3)$$

Nature does not tell Lab that value.

3. Nature determines the *unexplained expression level y_i*, which is the change in gene expression level due to factors other than the disease, that is, the change cannot be explained by

the disease. This is done by adding up the outcomes of three dice of six sides each (3d6) and applying this formula:

$$y_i = (3d6 - 10) \times \frac{\sigma}{3}. \tag{4.4}$$

Nature does not tell Lab that value.

4. Nature uses the value of μ_i from Step 2 and the value of y_i from Step 3 to determine $\bar{x}_i$, the observed expression level:

$$\bar{x}_i = \mu_i + y_i. \tag{4.5}$$

Nature tells Lab the value of $\bar{x}_i$.

4.3.3 Estimation of Effect Sizes

With the information supplied by Nature, Lab guesses the value of each effect size. The best guesses are the estimates computed by following this recipe:

1. The *grand mean* of the expression level is defined as

$$\bar{\bar{x}} = \frac{\bar{x}_1 + \cdots + \bar{x}_d}{d}.$$

It is the best estimate of *m*.

2. The *sample variance* is computed as

$$\hat{S}^2(x) = \frac{1}{d}\left(\left(\bar{x}_1 - \bar{\bar{x}}\right)^2 + \cdots + \left(\bar{x}_d - \bar{\bar{x}}\right)^2 \right) \tag{4.6}$$

to estimate S^2, the variance or squared amount of variability between the features. (Since a variance is just a standard deviation squared, the best estimate of the standard deviation would be $\sqrt{\hat{S}^2(x)}$, but that is not needed for the game.)

3. An empirical Bayes value of the *degree of shrinkage* is

$$B(x) = \frac{d-3}{d}\frac{\sigma^2}{\hat{S}^2(x)}, \tag{4.7}$$

which will be used in the next step to "shrink" the estimate of μ_i away from the value of $\bar{x}_i$ given by Nature and toward $\bar{\bar{x}}$, the grand mean computed in Step 1. (Notice that there is more shrinkage when σ^2, the within-feature variability, is large compared to $\hat{S}^2(x)$, the estimated between-feature variability.)

4. Finally, each effect size μ_i may be "estimated" or, more precisely, predicted by

$$\hat{\mu}_i(x) = (1 - B(x))\,\bar{x}_i + B(x)\,\bar{\bar{x}}.$$

That is Lab's best move. The estimate often used is $\bar{x}_i$, but it is less reliable in the sense that it will get a worse score on average.

4.3.4 Scoring

The error for gene i is $\hat{\mu}_i(x) - \mu_i$, the estimated effect size (from Lab) minus the true effect size (from Nature). The squared error for the ith gene is $\left(\hat{\mu}_i(x) - \mu_i\right)^2$. The sum of the squared errors (SSE) is computed by adding up the squared errors over all d genes:

$$\text{SSE} = \left(\hat{\mu}_1(x) - \mu_1\right)^2 + \left(\hat{\mu}_2(x) - \mu_2\right)^2 + \cdots + \left(\hat{\mu}_d(x) - \mu_d\right)^2, \tag{4.8}$$

which is how many points that Nature gets.

An easier but less accurate way to estimate μ_i is to just use $\bar{x}_i$ instead of $\hat{\mu}_i$. Its sum of squared errors is

$$\text{SSE}_{\text{naive}} = (\bar{x}_1(x) - \mu_1)^2 + (\bar{x}_2(x) - \mu_2)^2 + \cdots + (\bar{x}_d(x) - \mu_d)^2.$$

In most plays of the game and in most real studies, $\text{SSE} < \text{SSE}_{\text{naive}}$, making $\hat{\mu}_i$ well worth the extra computational effort. However, $\bar{x}_i$ can achieve lower error if d is very small, as explained in Section 5.4.2.

Players divide into Team A and Team B and stay on the same teams for two rounds of E&E. In the first round, Team A is Nature and Team B is Lab, but in the second round, Team B is Nature and Team A is Lab. The SSE made by Lab is counted as points for whichever team is Nature. The winner is the team that scored more points when it was Nature than when the other team was Nature. That means that Team A wins only if the SSE is higher in the first round than in the second round.

It is most efficient in practice to play both rounds at the same time. First, Team A and Team B generate data as Nature. Then, each team, as Lab, estimates the effect sizes generated by the other team.

4.4 UNDER THE HOOD: NORMAL DISTRIBUTIONS

This section explains the importance of the 3d6 in Advanced DE (§4.2.2) and in E&E (§4.3). The prerequisites are playing those games and working out Exercise G2 and Exercise G3 in Section 4.6.

4.4.1 Advanced DE's Normal Distributions

This subsection presents a simple way to generate random numbers from a normal distribution by rolling dice.

The random number 3d6, the sum of three six-sided dice, is approximately normal with a mean of 10 and a standard deviation of 3. That sentence is abbreviated by

$$3d6 \stackrel{.}{\sim} N\left(10, 3^2\right) \tag{4.9}$$

in the notation for normal distributions that was presented in Section 1.2. This normal distribution will be used to relate 3d6 to the normal distributions relevant to the ith gene ($i = 1, \ldots, d$).

In Advanced DE, Nature's simulations are based on the assumption that the average expression level $\bar{x}_i$, that is, mean $\log_2$ (sick/control), has a normal distribution of standard deviation $\sigma_i = 3s_i/\sqrt{n}$ and a mean of 0 if $A_i = 0$ or $\pm 3s_i$ if $A_i = 1$. In other words, the average expression levels have an approximate mean of 0 for the genes that are equivalently expressed but an approximate mean of $-3s_i$

or $+3s_i$ for the genes that are differentially expressed. In short,

$$
\bar{x}_i =
\begin{cases}
\left(\frac{3d6-10}{3}\right)\left(\frac{3s_i}{\sqrt{n}}\right) \sim N\left(0, \left(\frac{3s_i}{\sqrt{n}}\right)^2\right) & \text{if } A_i = 0 \\[2ex]
\left(\frac{3d6-10}{3}\right)\left(\frac{3s_i}{\sqrt{n}}\right) \pm 3s_i \sim N\left(\pm 3s_i, \left(\frac{3s_i}{\sqrt{n}}\right)^2\right) & \text{if } A_i = 1.
\end{cases}
$$

That formula emphasizes the special nature of the random number

$$
\frac{3d6 - 10}{3}.
$$

Formula (4.9) can be rearranged to give the approximate distribution of the *rescaled* average expression levels used by Lab in Advanced DE (§4.2.2):

$$
z_i = \frac{\bar{x}_i}{\sigma_i} = \frac{\bar{x}_i}{3s_i/\sqrt{n}} =
\begin{cases}
\frac{3d6-10}{3} \sim N(0,1) & \text{if } A_i = 0 \\[2ex]
\frac{3d6-10}{3} \pm \sqrt{n} \sim N\left(\pm\sqrt{n}, 1\right) & \text{if } A_i = 1.
\end{cases}
$$

That equation says the rescaled average expression levels have an approximate mean of 0 for the genes that are equivalently expressed but an approximate mean of $-\sqrt{n}$ or $+\sqrt{n}$ for the genes that are differentially expressed. The standard deviation of z_i is always about 1 since z_i was rescaled according to equation (4.2). To summarize, z_i is approximately distributed as $N(0,1)$, $N\left(\pm - \sqrt{n}, 1\right)$, or $N\left(+\sqrt{n}, 1\right)$.

What do those three normal distributions look like? Figure 4.6 displays histograms that approximate the probability density function of $N(0,1)$, whereas Figure 4.7 displays histograms that approximate the probability density functions of $N(-\sqrt{4}, 1)$ and $N(+\sqrt{4}, 1)$ (see Exercise G2). The exact probability density functions of all three distributions are displayed in Figure 4.8.

4.4.2 E&E's Normal Distributions

The strategy of Section 4.3.3 describes the best strategy of Lab in E&E to the extent that the assumptions of this subsection hold.

Assume that each component of the data

$$X_i = \langle X_{i1}, \ldots, X_{in} \rangle$$

of the ith gene or other feature is independent of the other components of that feature and that the sample mean

$$\bar{X}_i = \frac{X_{i1} + \cdots + X_{in}}{n}$$

follows this normal prior distribution of random mean μ_i and known standard error σ:

$$\bar{X}_i \sim N\left(\mu_i, \sigma^2\right), \quad i = 1, \ldots, d.$$

Then, in the case of Example 5.5 with the logarithms of base 10, the gene expression *fold change* is $10^{|\mu_i|}$, with the implication that any estimate of μ_i corresponds to an estimate of the fold change. Suppose further that the population means for the ith feature are independent and that each follows a common normal distribution:

$$\mu_i \sim N\left(m, S^2\right), \quad i = 1, \ldots, d,$$

where the mean m and standard deviation S are unknown, fixed parameters.

The *estimated posterior distribution* of μ_i is the normal distribution of mean $\hat{\mu}_i(x)$ and variance $(1 - B(x)) \times \sigma^2$, that is,

$$\left(\mu_i | \bar{X}_i = \bar{x}_i\right) \overset{\cdot}{\sim} N\left(\hat{\mu}_i(x), (1 - B(x)) \times \sigma^2\right), \quad i = 1, \ldots, d. \tag{4.10}$$

Recall the meanings of $B(x)$ and σ given in Section 4.3. Since the standard deviation is the square root of the variance, equation (4.10) means that the probability that the true value of the mean level of differential expression has about a 95% chance of being in the interval $\hat{\mu}_i(x) \pm 1.96 \times \sqrt{1 - B(x)} \times \sigma$:

$$P\left(\hat{\mu}_i(x) - 1.96 \times \sqrt{1 - B(x)} \times \sigma \leq \mu_i \leq \hat{\mu}_i(x) + 1.96 \right.$$
$$\left. \times \sqrt{1 - B(x)} \times \sigma \,\middle|\, \bar{X}_i = \bar{x}_i\right) \doteq 95\%. \tag{4.11}$$

The interval given by $\hat{\mu}_i(x) \pm 1.96 \times \sqrt{1 - B(x)} \times \sigma$ is an approximate *empirical Bayes 95% confidence interval* for μ_i.

4.5 BIBLIOGRAPHICAL NOTES

The parametric empirical Bayes method of Section 4.3.3 follows that of Morris (1983) as simplified by Ghosh et al. (2006, §§9.1, 9.2); see Carlin and Louis (2009, §5.2.1). Chapter 8 presents a simple parametric method of estimating the LFDR.

4.6 EXERCISES (C–E; G1–G4)

Exercise C. (a) Compute the 95% empirical Bayes confidence interval for μ_3 of Table 4.4. (b) Does the true value of μ_3 lie inside its confidence interval? Demonstrate the correctness of your answer.

Exercise D. (a) Fill in the blanks (_) of Table 4.1. (b) Put analogous notation ($x_{1,1}$, $x_{1,2}$, etc.) in the cells of Table 4.4.

Exercise E. (a) Demonstrate that the numbers in the $\bar{x}_i$ and $\hat{\mu}_i(x)$ columns of Table 4.4 are correct. (b) What do the numbers in the table mean biologically?

Exercise G1. Suppose that $P(1) = 10\%$ and $n = 10$. Nature follows the rules of Basic DE (§4.2.1) to simulate $\bar{x}_7 = -3$ and $p_7 = 0.03$ for gene #7. You, as a member of the Lab Team, used a computer to follow the same rules by rolling virtual dice. You set values of d to be large enough to generate Figures 4.1, 4.2, 4.3, and 4.4.

a. What is the prior *probability* of differential expression?

b. What is the prior *probability* of equivalent expression?

c. What is the prior *odds* of differential expression?

d. What is the likelihood ratio in favor of differential expression over equivalent expression? *Hint:* Some of the histograms are relevant.

TABLE 4.4 Gene Expression Results for Five Genes Using Logarithms of Base 10 and Using $\sigma = 1$

Gene	x_{i1}	x_{i2}	$\bar{x}_i$	$\hat{\mu}_i(x)$	μ_i	$P(\mu_i < -\log(2)|\bar{X}_i = \bar{x}_i)$	$P(\mu_i > \log(2)|\bar{X}_i = \bar{x}_i)$
$i = 1$	−1.9	−3.1	−2.5	−2.1	−1.2	96.0%	0.9%
$i = 2$	1.0	0.8	0.9	0.7	0.5	14.8%	67.1%
$i = 3$	1.4	0.3	0.8	0.7	0.9	16.8%	64.1%
$i = 4$	2.4	1.0	1.7	1.4	0.2	4.4%	86.4%
$i = 5$	0.1	−1.6	−0.8	−0.7	0.3	63.8%	17.0%

Note: The last two columns give the probability of under-expression and the probability of over-expression (treatment compared to control), respectively. The sum of those two columns gives $P(|\mu_i| > \log(2)|\bar{X}_i = \bar{x}_i)$, which is equal to $P(10^{|\mu_i|} > 2|\bar{X}_i = \bar{x}_i)$, the probability that the average expression changes by at least twofold (a factor of 2) in response to the treatment. All of these probabilities are based on equation (4.10). Here, $\hat{S}^2(x) = 2.2$. This table displays possible data for the experimental design of Example 3.1 but with only five genes measured instead of the 20,000 genes of the microarray. However, this table can also apply to non-microarray expression data involving substantially fewer genes (Example 5.5 of Section 5.4.1).

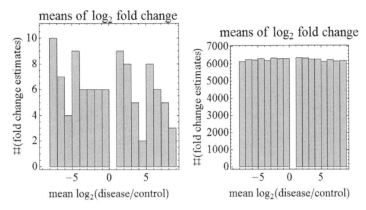

FIGURE 4.1 Histograms (approximate distributions) of logarithmic averaged expression ratios. There are $d = 100$ simulated genes on the left and $d = 100{,}000$ simulated genes on the right.

    e. What is the posterior *odds* of differential expression?

    f. What is the posterior *probability* of differential expression?

    g. What is the local false discovery rate?

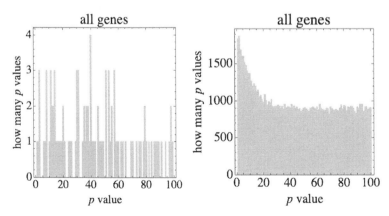

FIGURE 4.2 Histograms (approximate distributions) of p values as percentages for all genes of the computer simulation. There are $d = 100$ simulated genes on the left and $d = 100{,}000$ simulated genes on the right.

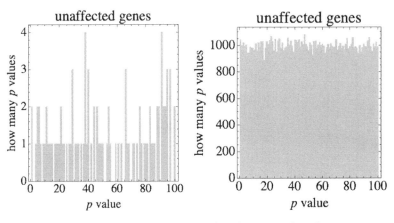

FIGURE 4.3 Histograms (approximate distributions) of p values as percentages for all genes *not* affected by the disease according to the computer simulation. There are $d = 100$ simulated genes on the left and $d = 100,000$ simulated genes on the right.

Exercise G2. Suppose that $P(1) = 15\%$ and $n = 4$. Nature follows the rules of Advanced DE (§4.2.2) to simulate gene #7, resulting in $\bar{x}_7 = -3$ and $s_7 = 1/2$. You, as a member of the Lab Team, used

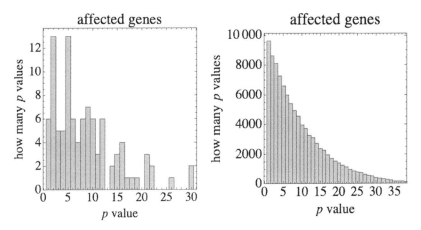

FIGURE 4.4 Histograms (approximate distributions) of p values as percentages for all genes affected by the disease according to the computer simulation. There are $d = 100$ simulated genes on the left and $d = 100,000$ simulated genes on the right.

a computer to follow the same rules by rolling virtual dice. You set values of d to be large enough to generate Figures 4.5, 4.6, and 4.7.

(a) What is the prior *probability* of differential expression?

(b) What is the prior *probability* of equivalent expression?

(c) What is the prior *odds* of differential expression?

(d) What is the likelihood ratio in favor of differential expression over equivalent expression?

(e) What is the posterior *odds* of differential expression?

(f) What is the posterior *probability* of differential expression?

(g) What is the local false discovery rate?

Exercise G3. (a) Play E&E (§4.3) using $d = 5$, first as Nature, then as Lab. What values do you get for $\bar{x}_i$, $\hat{\mu}_i(x)$, and μ_i for each gene ($i = 1, 2, 3, 4, 5$)? *Hint:* Modify the relevant columns of Table 4.4. (b) What are the SSE and $\text{SSE}_{\text{naive}}$ of that game? Was $\bar{x}_i$ or $\hat{\mu}_i(x)$

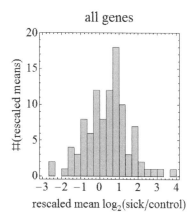

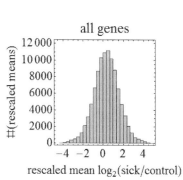

FIGURE 4.5 Histograms (approximate distributions) of rescaled means for all genes of the computer simulation. There are $d = 100$ simulated genes on the left and $d = 100,000$ simulated genes on the right.

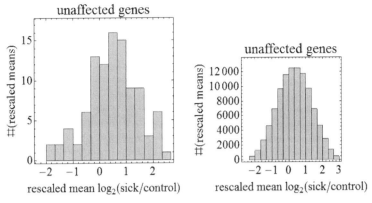

FIGURE 4.6 Histograms of rescaled means for all genes *not* affected by the disease according to the computer simulation. There are $d = 100$ simulated genes on the left and $d = 100,000$ simulated genes on the right. These histograms approximate the probability density function of a normal distribution with a mean of 0 (§4.4.1), as seen in the left panel of Figure 4.8.

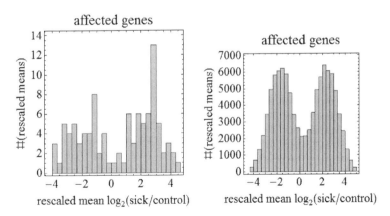

FIGURE 4.7 Histograms of rescaled means for all genes affected by the disease according to the computer simulation. There are $d = 100$ simulated genes on the left and $d = 100,000$ simulated genes on the right. These histograms approximate the probability density functions of normal distributions with means $-\sqrt{4}$ and $+\sqrt{4}$ (§4.4.1), as seen in the right panel of Figure 4.8.

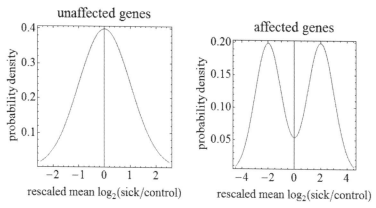

FIGURE 4.8 Probability density functions of normal distributions consistent with $n = 4$ human or animal subjects per sick or control group. The normal distribution on the left plot has a mean of 0 (equivalent expression); compare Figure 4.6. The two normal distributions on the right plot have means of $-\sqrt{4}$ (decrease in expression) and $+\sqrt{4}$ (increase in expression); compare Figure 4.7. The standard deviation of each of these distributions is 1, as required by the definition of a rescaled mean (§4.4.1).

better as an estimate of μ_i, the mean level of expression in gene i? (c) Answer Exercise G3(b) for Table 4.4 as it was before you played the game.

Exercise G4. (a) Which random numbers in E&E have approximately normal distributions? (b) What is the approximate normal distribution of each of those random numbers? *Hint 1:* The mean and standard deviation of any normal distribution distinguish it from all other normal distributions. Thus, Exercise G4(b) can be answered by stating the mean and standard deviation of each of the random numbers identified in Exercise G4(a). *Hint 2:* Understand Section 4.4.1, not only Section 4.4.2.

CHAPTER 5

Variations in Dimension and Data

It was obvious to me that my companion's mind was now made up about the case, although what his conclusions were was more than I could even dimly imagine. Several times during our homeward journey I endeavoured to sound him upon the point, but he always glided away to some other topic, until at last I gave it over in despair. It was not yet three when we found ourselves in our rooms once more. He hurried to his chamber and was down again in a few minutes dressed as a common loafer. With his collar turned up, his shiny, seedy coat, his red cravat, and his worn boots, he was a perfect sample of the class.

5.1 INTRODUCTION

The methodology covered for analyzing gene expression levels will now be extended to non-microarray data:

1. Two types of data from high-dimensional (high-d) genetics are considered in Section 5.2.

2. Various types of data of lower dimension are considered in Sections 5.3 and 5.4.

Much of the terminology of Chapter 3 and Section 4.3 is retained here. For example, *LFDR* refers to the local false discovery rate. However, since there is a possibility of missing data, the sample size n is no longer the number of measurements per feature because that number may possibly be less than n for some features, as in Example 5.6. However, n still represents the number of *biological replicates*, each of which consists of one or more distinct organisms. (If there is more than one measurement per biological replicate, each measurement is called a *technical replicate* since it can be used to assess measurement error as opposed to biological variation. Measurements should be averaged within each biological replicate over all technical replicates before applying the following estimators or other higher-level statistical methods.)

5.2 HIGH-DIMENSIONAL GENETICS

The method of LFDR estimation of Section 3.4 can be applied to high-d genetics data with the understanding that speaking of features as affected or unaffected is no longer appropriate, as the following examples make clear.

EXAMPLE 5.1: GENOME-WIDE ASSOCIATION STUDIES

A *genome-wide association* (GWA) study is a genetic association study (Example 1.4) in which the genotypes of hundreds of thousands of SNPs are measured. A contingency table like Table 1.4 can be set up for each SNP, resulting in hundreds of thousands of contingency tables.

In a typical GWA study, there is a measurement of which nucleotide occurs at a specific DNA site for each human subject in a disease or *case* group and for each human subject in a healthy or *control* group for each of d DNA sites, where $d > 10^5$. The occurrence of appreciable variation in nucleotides (and thus alleles) within a population of organisms at a specific DNA site is called a single-nucleotide polymorphism

(SNP). In terms of the notation of Chapter 3, the n-tuple of nucleotide measurements at the ith site (or SNP) is denoted by

$$x_i \text{ for } i = 1, \ldots, d$$

and the p value $p\,(x_i)$ is derived from a statistical test of the null hypothesis that the ith SNP is not associated with the disease. (Section 1.4 shows a simple way to get a p value.) Since d is very large, the LFDR estimation method of Section 3.4 can be applied, in which case a discovery is of disease-SNP association rather than of differential gene expression. Thus, if the estimate of the LFDR is small, the appropriate inference would be that SNP i affects the disease status; this differs from the previous interpretation of a feature being affected by a treatment. In other words, the previous terminology is inappropriate here since while a human subject's genes might influence whether she contracts a disease, whether she contracts a disease cannot possibly have any effect on her genes.

See http://www.genome.gov/20019523 for more information on GWA studies, and see Example 1.4 on more general genetic association studies.

EXAMPLE 5.2: QTL MAPPING STUDIES

In *quantitative-trait loci* (QTL) mapping studies, *genotype* data on alleles at locations in the genome called *loci* and *phenotype* data on quantitative (continuous) traits such as weight are used to test for each locus the null hypothesis that it is not linked to the locus of a gene that influences the phenotype. Two loci are said to be *linked* if inheriting an allele at one locus happens more frequently if an allele at a linked locus is inherited. If there are a large number of p values each corresponding to a different locus, the LFDR estimation method of Section 3.4 can be applied. A QTL is a locus linked to a gene that influences one or more of the quantitative traits measured. A discovery of linkage is a discovery of a QTL or, in the statistical genetics literature, a *putative QTL*. As in Example 5.1, the genotype affects the phenotype, not vice versa, so a discovery of a QTL suggests linkage to a gene that affects the quantitative trait.

5.3 SUBCLASSES AND SUPERCLASSES

A $\tilde{d}$-tuple of p values

$$\tilde{\mathbf{p}} = \langle p(x_1), \ldots, p(x_{\tilde{d}}) \rangle$$

is called a *class* of p values, and the two classes of p values

$$\mathbf{p}' = \langle p(x_1'), \ldots, p(x_{d'}') \rangle, \tag{5.1}$$
$$\mathbf{p}'' = \langle p(x_1''), \ldots, p(x_{d''}'') \rangle$$

are called *subclasses* of $\tilde{\mathbf{p}}$ if

$$\tilde{\mathbf{p}} = \langle \mathbf{p}', \mathbf{p}'', \ldots \rangle.$$

If $\mathbf{p}'$ is a subclass of $\tilde{\mathbf{p}}$, then $\tilde{\mathbf{p}}$ is a *superclass* of $\mathbf{p}'$. Additional terms will be introduced as needed.

Continuing Example 3.2, the following examples of gene expression data analysis illustrate the need for this section's methods of determining whether to separate analysis of a subclass from analysis of a superclass, i.e., whether to use $d = d'$ or $d = \tilde{d}$ for the purpose of estimating local false discovery rates. This is an instance of what philosophers call the problem of the *reference class*, the class used to compute a probability such as that approximated by the LFDR.

EXAMPLE 5.3: BIOCHEMICAL PATHWAY

A p value $p(x_i)$ for testing the null hypothesis of equivalent expression of the ith gene is available for each of $\tilde{d}$ genes.

However, the biologist is particularly interested in the $d' \gg 1$ genes known to play a role in a particular biochemical pathway. Whether the subclass of genes in the pathway or the superclass of all genes is used can have a large effect on the NFDR, as seen in Table 3.1. The choice of reference class similarly affects the LFDR.

Judging that affected genes in the biochemical pathway will have a different distribution of expression levels than affected genes in other biochemical pathways, the statistician collaborating with the biologist separately estimates the LFDR for each subclass. Thus, in terms of the equations given below, the data are analyzed using equation (5.3) rather than equation (5.2). However, without knowledge that would distinguish the expression distributions of different affected genes within the pathway, the LFDR is estimated using the entire superclass.

EXAMPLE 5.4: MULTIPLE COMPARISONS

The expression levels of d' genes are measured before and after application of each of two different perturbations: a high-calorie diet and a restricted-calorie diet; the control group of mice have a moderate-calorie diet. Let x, y, and z denote the log-transformed expression levels of the low-, moderate-, and high-calorie groups, respectively. Further, let

$$\mathbf{p}' = \langle p\left(x_1, y_1\right), \ldots, p\left(x_{d'}, y_{d'}\right)\rangle$$

denote the d' p values for testing equivalent expression between the low-calorie population and the control population,

$$\mathbf{p}'' = \langle p\left(z_1, y_1\right), \ldots, p\left(z_{d'}, y_{d'}\right)\rangle$$

the d' p values for testing equivalent expression between the high-calorie population and the control population, and

$$\mathbf{p}''' = \langle p\left(z_1, x_1\right), \ldots, p\left(z_{d'}, x_{d'}\right)\rangle$$

the d' p values for testing equivalent expression between the high-calorie population and the low-calorie population. The superclass of each of those three subclasses is

$$\tilde{\mathbf{p}} = \langle p\left(x_1, y_1\right), \ldots, p\left(x_{d'}, y_{d'}\right), p\left(z_1, y_1\right), \ldots, \\ p\left(z_{d'}, y_{d'}\right), p\left(z_1, x_1\right), \ldots, p\left(z_{d'}, x_{d'}\right) \rangle,$$

which consists of

$$\tilde{d} = 3d'$$

p values.

These examples are summarized in Table 5.1.

The LFDR of the ith feature relative to a superclass is

$$\tilde{\ell}\left(x_i\right) = \frac{\left(\tilde{d}_0/\tilde{d}\right) g_0\left(p\left(x_i\right)\right)}{\left(\tilde{d}_0/\tilde{d}\right) g_0\left(p\left(x_i\right)\right) + \left(1 - \left(\tilde{d}_0/\tilde{d}\right)\right) \tilde{g}_1\left(p\left(x_i\right)\right)},$$
$$i = 1, \ldots, \tilde{d}, \tag{5.2}$$

where, for that superclass, $\tilde{g}_1$ and $\tilde{g}_0$ are the density function of the data generated by the affected features and the proportion of unaffected features, respectively. Likewise, the LFDR of the ith feature relative to the first subclass of that superclass is

$$\ell'\left(x_i\right) = \frac{\left(d'_0/d'\right) g_0\left(p\left(x_i\right)\right)}{\left(d'_0/d'\right) g_0\left(p\left(x_i\right)\right) + \left(1 - \left(d'_0/d'\right)\right) g'_1\left(p\left(x_i\right)\right)},$$
$$i = 1, \ldots, d', \tag{5.3}$$

where, for that subclass, g'_1 and π'_0 are the density function of the data generated by the affected features and the proportion of unaffected features, respectively. The main issue addressed in

TABLE 5.1   Summary of Examples 5.3 and 5.4

Subclass	Superclass	d'	$\tilde{d}$
A pathway's genes	All genes	Number of genes in pathway	Number of genes
A comparison's p values	All p values	Number of genes	Number of p values

this section is whether $\tilde{\ell}(x_i)$ or $\ell'(x_i)$ is the more relevant LFDR associated with the ith feature.

If affected genes for a particular comparison between two groups will have a different distribution of expression levels than affected genes for comparisons between other groups, the data are analyzed using three separate applications of equation (5.3) rather than one application of equation (5.2).

5.4 MEDIUM NUMBER OF FEATURES

Section 5.3 shows how to analyze data of a reduced dimension using a method that applies to large dimension. When the dimension is too small for methods designed for high-dimension data, special methods are needed.

5.4.1 Empirical Bayes Methods

The best strategies of Lab in the DE games and E&E (Chapter 4) are examples of *parametric empirical Bayes* methods. A recipe of statistical data analysis is called an *empirical Bayes* method if it involves the computation of a Bayesian posterior distribution (§3.2.1) on the basis of an *estimated* prior distribution, as discussed further in Section 3.2.2. The empirical Bayes method of estimating the LFDR of Section 3.4 is considered *semi-parametric* rather than fully parametric since its only unknown parameters are d_0 and any parameters involved in the statistical test used to generate the p values. In that method, the probability density function g is estimated completely *nonparametrically* since it does not rely on any error model determined by parameters. Consequently, it is only reliable for large numbers of features, as reflected in the requirement that $1 \ll \delta \ll d$.

If the number of features is medium, i.e.,

$$3 \ll d \lesssim 1000,$$

as in the examples of Section 5.4.2, then the *parametric empirical Bayes* (PEB) methods of Sections 4.6 and 4.3.3 will tend to be more reliable since each leverages information in a specific error model determined by parameters, provided that a distribution in the error model is sufficiently close to the data-generating process. Whereas biological data-generating processes are experiments or observational studies, artificial data-generating processes are simulations such as those of Nature in the games of Chapter 4.

EXAMPLE 5.5: GENE EXPRESSION STUDIES

While previous examples used the microarray as the instrument for measuring gene expression, this example instead uses reverse transcription polymerase chain reaction (RT-PCR) instruments to measuring gene expression. Typically, the expression of many fewer genes are measured than is possible with microarrays, and it is not unusual to have expression measurements of only between 10 and 30 genes. Thus, all of the previous gene expression examples are relevant to RT-PCR data except that the number of genes is much smaller than with microarrays, which might simultaneously measure 5000 or more genes. RT-PCR is often used to confirm genes discovered to be differentially expressed on the basis of microarray data.

The number i will stand for the label of the gene, and the number j will stand for the label of the biological replicate. As with a simple microarray experiment, the random number X_{ij} could be the observable, log-transformed difference in levels of expression of gene i between the jth individual of the treatment group and the jth individual of the control group. In a simple parametric error model, X_{ij} is a normally distributed random number of unknown mean μ_i and unknown variance σ^2, a measure of the variability within each feature. The outcome of X_{ij} is a number such as 2.32. The lowercase x_{ij} will stand for any such number that is observed as an outcome of the random X_{ij}. Each component x_{ij} of the observed n-tuple

$$x_i = \langle x_{i1}, \ldots, x_{in} \rangle$$

could be the logarithm of a treatment-to-control expression ratio for the ith gene such as the ratio of the expression in a treatment individual to that of a control individual, as in Table 1.1. Alternatively, each component could correspond to a different treatment individual compared to a common control reference.

Table 4.4 shows solutions found using parametric empirical Bayes methods.

5.4.2 Applications to Data of Medium Dimension

While PEB methods can be applied to genome-scale data $(d \gtrsim 1000)$, they can also be applied when the number of features d is too small for the use of $semi$-parametric empirical Bayes methods such as the LFDR estimator of Section 3.4. When data are generated by a medium-d measuring process such as RT-PCR protocols (Example 5.5) or by mass spectroscopy (Example 5.6), the number of features measured is too low to reliably estimate the LFDR with semi-parametric methods. In these medium-d situations, the data may be more reliably analyzed using a parametric method such as that of Section 4.3.3.

EXAMPLE 5.6: PROTEOMICS AND METABOLOMICS STUDIES

In addition to measuring the amount of mRNA (gene product) in a cell, biologists also measure the relative concentration levels of other features in the cell such as proteins and metabolites. However, since such abundance levels are often measured using mass spectroscopy, not all of the proteins or metabolites necessarily have measurements for all individuals in the treatment and control groups. Then, given n biological replicates, every x_i is an n_i-tuple, where $n_i = 1, \ldots, n$ is the number of abundance-level measurements available for the ith feature. For any $i = 1, \ldots, d$ such that

$$n_i < n,$$

feature i is said to be *missing data*. Ignoring data for features with $n_i = 1$ and setting d to the number of features for which $n_i \geq 2$, the method of Section 4.3.3 can be applied separately to each of the subclasses $\mathbf{x}_2, \mathbf{x}_3, \ldots, \mathbf{x}_n$ defined such that

$$d = d(2) + \cdots + d(n).$$

See Section 5.3 for more information on superclasses and subclasses.

To illustrate how the data may be analyzed using a parametric empirical Bayes method, the method of Section 4.3.3 is now adapted to the reference class problem of Section 5.3. Instead of $\tilde{d}$-tuples, d'-tuples, d''-tuples, etc. of p values, we have tuples of observations. In particular, a $\tilde{d}$-tuple of observations

$$\tilde{\mathbf{x}} = \left\langle x_1, \ldots, x_{\tilde{d}} \right\rangle$$

is called a *superclass* of observations, and the two classes of observations

$$\mathbf{x}' = \left\langle x_1', \ldots, x_{d'}' \right\rangle, \tag{5.4}$$

$$\mathbf{x}'' = \left\langle x_1'', \ldots, x_{d''}'' \right\rangle \tag{5.5}$$

are called *subclasses* of $\tilde{\mathbf{x}}$ if

$$\tilde{\mathbf{x}} = \left\langle \mathbf{x}', \mathbf{x}'', \ldots \right\rangle.$$

The quantities of Section 4.3.3 are replaced with

$$\bar{\bar{\mathbf{x}}} = \frac{\bar{x}_1 + \cdots + \bar{x}_{\tilde{d}}}{\tilde{d}},$$

$$\hat{S}^2(\tilde{\mathbf{x}}) = \frac{1}{\tilde{d}} \left(\left(\bar{x}_1 - \bar{\bar{\mathbf{x}}} \right)^2 + \cdots + \left(\bar{x}_{\tilde{d}} - \bar{\bar{\mathbf{x}}} \right)^2 \right),$$

$$B\left(\tilde{\mathbf{x}}\right) = \frac{\tilde{d} - 3}{\tilde{d}} \frac{\sigma^2}{\hat{S}^2\left(\tilde{\mathbf{x}}\right)},$$

$$\hat{\mu}_i\left(\tilde{\mathbf{x}}\right) = \left(1 - B\left(\tilde{\mathbf{x}}\right)\right) \bar{x}_i + B\left(\tilde{\mathbf{x}}\right) \bar{\bar{\mathbf{x}}}$$

if the superclass is used or, assuming $d' \geq 4$, with

$$\overline{\overline{\mathbf{x}'}} = \frac{\bar{x}'_1 + \cdots + \bar{x}'_{d'}}{d'},$$

$$\hat{S}^2\left(\mathbf{x}'\right) = \frac{1}{d'} \left(\left(\bar{x}'_1 - \overline{\overline{\mathbf{x}'}}\right)^2 + \cdots + \left(\bar{x}'_{d'} - \overline{\overline{\mathbf{x}'}}\right)^2 \right),$$

$$B\left(\mathbf{x}'\right) = \frac{d' - 3}{d'} \frac{\sigma^2}{\hat{S}^2\left(\mathbf{x}'\right)},$$

$$\hat{\mu}_i\left(\mathbf{x}'\right) = \left(1 - B\left(\mathbf{x}'\right)\right) \bar{x}'_i + B\left(\mathbf{x}'\right) \overline{\overline{\mathbf{x}'}}$$

if the first subclass is used instead. If $d' \leq 3$, one could resort to the naive estimate instead of using the superclass:

$$\hat{\mu}_i\left(\mathbf{x}'\right) = \bar{x}'_i$$

The same pattern holds for the other subclasses.

EXAMPLE 5.7: A LIPID STUDY

A *lipid* is a metabolite made up of fatty acids. The simultaneous study of several lipids is the driving force behind the emerging field of *lipidomics*. Imagine that a biologist measures the levels of 40 lipids before and after application of each of the two different perturbations described in Example 5.4. Such lipids include 15 phospholipids and 25 steroids. For each of the three diet comparisons and for each of the 40 lipids, there are three measured log-transformed differences in concentration between diet levels, except that there are only two such measurements for each of 11 steroids for the high-calorie-to-control comparison. Since phospholipids and steroids have different chemical structures and different functions, their concentration levels are not modeled as constituting a single exchangeable sequence, even within the same diet comparison and number of measurements. As argued in Example 5.4, the comparisons must be separated before data analysis, and as argued in Example 5.6, data of different numbers of measurements must also be separated before analysis.

Thus, for each of the seven cells of Table 5.2 that has a shrinkage degree rather than "N/A," different values of $\bar{\bar{x}}$, $B(x)$, and $\hat{S}^2(x)$ are computed for the purposes of predicting each lipid's mean level of log-transformed concentration difference between two diets and estimating the posterior distribution of that mean. In the table, Roman-numeric superscripts on the data n'-tuples take the place of the primes $(')$ on p value tuples used in equations (5.4) –(5.5) to label subclasses, whereas the numbers before the arrows denote the numbers of features used as d for the computations of Section 4.3.3. For instance, in the high-calorie vs. control comparison for steroids, equation (4.7) is replaced with

$$B\left(\mathbf{x}_3^{\mathrm{V}}\right) = \frac{14 - 3}{14} \frac{\sigma^2}{\hat{S}^2\left(\mathbf{x}_3^{\mathrm{V}}\right)}. \tag{5.6}$$

Exercise G5 clarifies this example.

The advantage of using a subclass or naive estimator as $\hat{\mu}_i(x)$ is that each tends to have a bias smaller than the superclass, whereas the disadvantage is that the subclasses and naive estimators tend to have higher variances than the superclass:

$$\mathrm{bias} = \frac{\left(\hat{\mu}_1(x) - \mu_1\right) + \left(\hat{\mu}_2(x) - \mu_2\right) + \cdots + \left(\hat{\mu}_d(x) - \mu_d\right)}{d},$$

TABLE 5.2 Complex Subclasses (Example 5.7)

	Phospholipids ($n' = 3$)	Steroids ($n' = 3$)	Steroids ($n' = 2$)
Low-calorie vs. control	$d(3) = 15 \longrightarrow B(\mathbf{x}^{I})$	$d(3) = 25 \longrightarrow B(\mathbf{x}^{IV})$	N/A
High-calorie vs. control	$d(3) = 15 \longrightarrow B(\mathbf{x}^{II})$	$d(3) = 14 \longrightarrow B(\mathbf{x}_3^{V})$	$d(2) = 11 \longrightarrow B(\mathbf{x}_2^{V})$
High-calorie vs. low-calorie	$d(3) = 15 \longrightarrow B(\mathbf{x}^{III})$	$d(3) = 25 \longrightarrow B(\mathbf{x}^{VI})$	N/A

Note: While the Roman numerals only go up to VI, there are a total of seven subclasses for seven separate computations of the shrinkage degree since the superclass $\mathbf{x}^{V}$ was broken into the subclasses $\mathbf{x}_3^{V}$ and $\mathbf{x}_2^{V}$ due to missing data. Here, n' is the number of measurements per lipid.

variance =

$$\frac{\left(\hat{\mu}_1(x) - \overline{\hat{\mu}}(x)\right)^2 + \left(\hat{\mu}_2(x) - \overline{\hat{\mu}}(x)\right)^2 + \cdots + \left(\hat{\mu}_d(x) - \overline{\hat{\mu}}(x)\right)^2}{d},$$

$$\overline{\hat{\mu}}(x) = \frac{\hat{\mu}_1(x) + \hat{\mu}_2(x) + \cdots + \hat{\mu}_d(x)}{d}.$$

The sum of the squared errors (SSE) defined by equation (4.8) can be expressed in bias and variance components:

$$\text{SSE} = d \times \left(\text{bias}^2 + \text{variance}\right).$$

An optimal approach is to choose the combination of the super-class estimators, subclass estimators, and naive estimators to balance the bias and the variance in such a way that the SSE is as low as possible. Note, however, that Lab cannot compute the bias from the data since $\mu_1, \mu_2, \ldots, \mu_d$ are unknown.

A rule of thumb is to use subclasses whenever they are large enough to achieve a reasonably low variance. When the subclass is too small to attain low variance for a feature, the superclass could be used for estimation of that feature's effect size. However, if the bias of the superclass is considered too high for a feature, then its subclass should be used if the subclass has at least four features, and the naive estimator should be used if it has only one to three features.

5.5 BIBLIOGRAPHICAL NOTES

Efron and Tibshirani (2002) emphasized the importance of exchangeability when estimating the LFDR. Subramanian et al. (2005), Newton et al. (2007), Efron and Tibshirani (2007), and Efron (2008) considered the problem of Example 5.3, and Bickel (2004) that of Example 5.4. New approaches to the reference class problem of LFDR estimation may be found in Karimnezhad and Bickel (2018) and Aghababazadeh et al. (2018).

The resource http://www.genome.gov/20019523 was accessed 30 May 2019.

5.6 EXERCISE (G5)

Exercise G5. In order to answer questions (a)–(d) at the end of this exercise, play the following variation of the E&E game described in Section 4.3. The game simplifies Example 5.7. Nature follows these steps:

1. Choose a single value of σ for all seven of the lipids (four phospholipids and three steroids). This value is a number between 0.1 and 4. Tell Lab the value of σ.

2. Choose two values of S, one value of S for the four phospholipids and another value of S for the three steroids. Each S value is a number between 0.1 and 4, but they cannot differ by more than a factor of 2, that is, neither value of S can be more than twice the other value of S. Do not tell Lab either value.

3. For each of the phospholipids ($i = 1, 2, 3, 4$), do the following:

    i. Use the phospholipid S and $m = 0$ in equation (4.3) to get μ_i, the mean change in the abundance level of lipid i.

    ii. Then use the phospholipid σ and $m = 0$ in equation (4.4) to get y_i, the unexplained variation in the abundance level of lipid i.

    iii. Use equation (4.5) to get $\bar{x}_i$, the observed mean abundance level of lipid i.

    iv. Tell Lab $\bar{x}_i$ but not μ_i or y_i.

4. For each of the steroids ($i = 5, 6, 7$), do the following:

    i. Use the steroid S and $m = 0$ in equation (4.3) to get μ_i, the mean change in the abundance level of lipid i.

    ii. Then use the steroid σ and $m = 0$ in equation (4.4) to get y_i, the unexplained variation in the abundance level of lipid i.

  iii. Use equation (4.5) to get $\bar{x}_i$, the observed mean abundance level of lipid i.

  iv. Tell Lab $\bar{x}_i$ but not μ_i or y_i.

On the basis of nature's moves, Lab answers these questions:

(a) What are the numeric values of $\bar{x}_1 \ldots, \bar{x}_{\bar{d}}$, $\bar{x}'_1 \ldots, \bar{x}'_{d'}$, and $\bar{x}''_1 \ldots, \bar{x}''_{d''}$ in the game?

(b) What are Lab's best estimates of $\mu_1, \ldots, \mu_7$?

(c) What are the biases, variances, and SSEs?

(d) What are the 95% confidence intervals? *Hint:* Apply the reasoning of Section 5.4.2 to equation (4.11).

CHAPTER 6

Correcting Bias in Estimates of the False Discovery Rate

My experience of camp life in Afghanistan had at least had the effect of making me a prompt and ready traveller. My wants were few and simple, so that in less than the time stated I was in a cab with my valise, rattling away to Paddington Station. Sherlock Holmes was pacing up and down the platform, his tall, gaunt figure made even gaunter and taller by his long grey travelling-cloak and close-fitting cloth cap.

"It is really very good of you to come, Watson," said he. "It makes a considerable difference to me, having someone with me on whom I can thoroughly rely. Local aid is always either worthless or else biassed. If you will keep the two corner seats I shall get the tickets."

BACKGROUND

Conventional methods of adjusting p values for multiple comparisons control a family-wise error rate (FWER) such as a genome-wise error rate. The recognition that they lead to excessive false negative rates in genomics applications has led to widespread use of false discovery rates (FDRs) in place of the conventional adjustments. While this is an improvement, the way FDRs are used in the analysis of genomics data leads to the opposite problem, excessive false positive rates. In this sense, the FDR overcorrects for the excessive conservatism (bias toward false negatives) of the FWER-controlling methods of adjusting p values.

6.1 WHY CORRECT THE BIAS IN ESTIMATES OF THE FALSE DISCOVERY RATE?

Your job is to provide the selection committee with a list of applicants predicted to graduate with honors if selected. While you are trying to decide on whether to report those with less than a 5%, 20%, or 50% probability of failure, Dr. F. D. Rate informs you of a popular new procedure. It replaces each applicant's failure probability with its average with the failure probabilities of all applicants with lower failure probabilities. Having decided to predict the graduation of each applicant with less than 20% failure probability, you now wonder how to calibrate the accurately estimated average failure probability among the better candidates in order to make it reasonable as an estimate of the applicant's failure probability.

Given a gene expression data set, you need to determine which genes to consider differentially expressed. If you decide that only genes with more than 80% estimated probability of differential expression belong on the list, then every gene on the list will have less than 20% estimated probability that its null hypothesis of equivalent expression is true. That probability is the local false discovery rate (LFDR). Since that probability is difficult to estimate accurately, the false discovery rate looks like a viable alternative. In fact, software is readily available to report the FDR achieved by each gene. However, since the FDR achieved by a gene is for practical purposes an average of the LFDRs over all genes with lower LFDRs of that gene, the achieved FDR is by construction too low to use as an estimate of the null hypothesis probability.

That bias requires a correction in order to construct the desired list of genes with over 80% estimated probability of differential expression.

The two bias corrections explained in this chapter transform achieved FDRs into LFDR estimates. As calibrated FDR values, those LFDR estimates in effect determine reasonable FDR thresholds to apply to the output of genomics software packages. This applicability of the bias corrections to popular FDR methods achieves the primary goal of this chapter: to make the benefits of LFDR estimation more accessible.

As the flaw in the procedure is practical, it may be most clearly seen in terms of particular applications to genomics data. Suppose that the null hypothesis is that a gene is not differentially expressed. Selecting a gene for further consideration on the basis of accurate FDR estimation in effect selects a gene on the basis of the probability that a more significant gene is differentially expressed. That is equivalent to selecting an applicant for admission on the basis of an average of his or her probability of failure with the failure probabilities of all the better applicants.

6.2 A MISLEADING ESTIMATOR OF THE FALSE DISCOVERY RATE

Recall from equation (3.5) that

$$\widehat{\text{FDR}}(\alpha) = \begin{cases} \dfrac{\alpha d}{\#(p(x_i) \leq \alpha)} & \text{if } \dfrac{\alpha d}{\#(p(x_i) \leq \alpha)} < 1 \\ 1 & \text{if } \dfrac{\alpha d}{\#(p(x_i) \leq \alpha)} > 1. \end{cases}$$

It is tempting to plug in each p value for α, yielding the *achieved FDR* estimate for test j:

$$\widehat{\text{FDR}}(p(x_j)) = \begin{cases} \dfrac{p(x_j)d}{\#(p(x_i) \leq p(x_j))} & \text{if } \dfrac{p(x_j)d}{\#(p(x_i) \leq p(x_j))} < 1 \\ 1 & \text{if } \dfrac{p(x_j)d}{\#(p(x_i) \leq p(x_j))} > 1. \end{cases}$$

However, that can be misleadingly low because, according to equation (3.10), the achieved FDR satisfies

$$\text{FDR}\left(p\left(x_j\right)\right)$$

$$\approx \frac{\text{LFDR}\left(p\left(x_1\right)\right) + \text{LFDR}\left(p\left(x_2\right)\right) + \cdots + \text{LFDR}\left(p_{\#\left(p\left(x_i\right)\leq p\left(x_j\right)\right)}\right)}{\#\left(p\left(x_i\right) \leq p\left(x_j\right)\right)} \tag{6.1}$$

$$= \frac{\text{LFDR}\left(p\left(x_1\right)\right) + \text{LFDR}\left(p\left(x_2\right)\right) + \cdots + \text{LFDR}\left(p\left(x_j\right)\right)}{\#\left(p\left(x_i\right) \leq p\left(x_j\right)\right)} \tag{6.2}$$

if the p values are sorted such that $p\left(x_1\right) < p\left(x_2\right) < \cdots < p\left(x_d\right)$. Assuming that lower p values correspond to lower LFDRs and that there are no duplicate p values, $\text{LFDR}\left(p\left(x_1\right)\right) < \text{LFDR}\left(p\left(x_2\right)\right) < \cdots < \text{LFDR}\left(p\left(x_d\right)\right)$. Since $\text{LFDR}\left(p\left(x_j\right)\right)$ is the highest LFDR in the numerator of equation (6.2), it is higher than the average of those LFDRs:

$$\text{LFDR}\left(p\left(x_j\right)\right)$$

$$> \frac{\text{LFDR}\left(p\left(x_1\right)\right) + \text{LFDR}\left(p\left(x_2\right)\right) + \cdots + \text{LFDR}\left(p\left(x_j\right)\right)}{\#\left(p\left(x_i\right) \leq p\left(x_j\right)\right)}$$

$$= \text{FDR}\left(p\left(x_j\right)\right).$$

In short, $\text{FDR}\left(p\left(x_j\right)\right) < \text{LFDR}\left(p\left(x_j\right)\right)$. That means that $\widehat{\text{FDR}}\left(p\left(x_j\right)\right)$ is biased to be too low to reliably estimate $\text{LFDR}\left(p\left(x_j\right)\right)$. The practical consequence is that too many genes will be considered differentially expressed since the probability of making false discoveries is higher than indicated by $\widehat{\text{FDR}}\left(p\left(x_j\right)\right)$.

While the local FDR of a null hypothesis is simply the posterior probability that the null hypothesis is true, the other concepts of this section are more complicated. The nonlocal FDR at a significance level is the probability that a randomly selected p value that is less than the significance level is true. The achieved FDR of a null hypothesis is the nonlocal FDR with the significance level set equal to the p value of that null hypothesis. That means that it is the probability that another null hypothesis is true given that its p value is less than the p value of the null hypothesis being considered. That is why the achieved FDR is less than the posterior probability that the null hypothesis is true. The extent of that discrepancy is a bias in need of correction.

6.3 CORRECTED AND RE-RANKED ESTIMATORS OF THE LOCAL FALSE DISCOVERY RATE

One solution of the problem described in Section 6.2 is to correct the bias in the $\widehat{FDR}(p(x_j))$. A much better estimate of LFDR $(p(x_j))$ would be the *corrected FDR* (CFDR),

$$
\text{CFDR}(x_j) = \begin{cases} \left(\frac{1}{1-1+1}\right)\widehat{FDR}(p(x_j)) & \text{if } j = 1 \\ \left(\frac{1}{2-1+1} + \frac{1}{2-2+1}\right)\widehat{FDR}(p(x_j)) & \text{if } j = 2 \\ \left(\frac{1}{3-1+1} + \frac{1}{3-2+1} + \frac{1}{3-3+1}\right)\widehat{FDR}(p(x_j)) & \text{if } j = 3 \\ \left(\frac{1}{j-1+1} + \frac{1}{j-2+1} + \cdots + \frac{1}{j-j+1}\right)\widehat{FDR}(p(x_j)) & \text{if } j \geq 4. \end{cases}
$$
(6.3)

A related solution is to change the ranks of $\widehat{FDR}(p(x_j))$, the achieved estimated FDR. It results in the *re-ranked false discovery rate* (RFDR) as an estimator of LFDR $(p(x_j))$, the local false discovery rate of the jth smallest p value. It is defined by

$$
\text{RFDR}(x_j) = \begin{cases} \widehat{FDR}\left(p\left(x_{[1.6j]}\right)\right) & \text{if } [1.6j] \leq d \\ 1 & \text{if } [1.6j] > d, \end{cases}
$$
(6.4)

where $[1.6j]$ is $1.6j$ rounded off to the closest whole number. That means that RFDR (x_j) replaces the jth smallest achieved estimated FDR with the $[1.6j]$th smallest achieved estimated FDR. For example, for $j = 10$, the tenth smallest achieved estimated FDR is replaced with the 16th smallest achieved estimated FDR since $[1.6 \times 10] = [16] = 16$, assuming $d \geq 16$. Likewise, for $j = 1$, the very smallest achieved estimated FDR is replaced with the second smallest achieved estimated FDR since $[1.6 \times 1] = [1.6] = 2$, assuming $d \geq 2$.

6.4 APPLICATION TO GENE EXPRESSION DATA ANALYSIS

Figure 6.1 reveals that $\widehat{\text{FDR}}\left(p\left(x_j\right)\right)$ is much less than CFDR $\left(x_j\right)$ and RFDR $\left(x_j\right)$ for the 545 genes with $\widehat{\text{FDR}}\left(p\left(x_j\right)\right) \leq 0.2$. By comparison, only 69 or 344 of which satisfy CFDR $\left(x_j\right) \leq 0.2$ or RFDR $\left(x_j\right) \leq 0.2$, respectively. Lowering the threshold defining what is meant by a "discovery" has no effect on the relative magnitudes of the discrepancies (Figure 6.2). Since the number of discoveries in Figure 6.2 is the sum of the number of true discoveries and the number of false discoveries, higher numbers of discoveries tend to coincide with higher numbers of false discoveries.

6.5 BIBLIOGRAPHICAL NOTES

Hong et al. (2009, Fig. 3) and Bickel (2012b) warned against interpreting quantities like $\widehat{\text{FDR}}\left(p\left(x_j\right)\right)$ as LFDR estimators. Bickel and Rahal (2019) introduced the calibrated FDR and the re-ranked FDR and then applied them to gene expression data analysis as per Section 6.4. The marked discrepancy between LFDR and NFDR

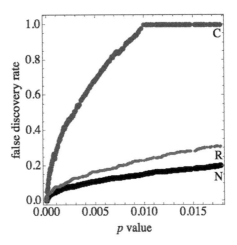

FIGURE 6.1 Local ("C"; "R") and nonlocal ("N") false discovery rate estimates as a function of $p\left(x_j\right)$ for gene expression data: "N" = nonlocal FDR; "C" = CFDR; "R" = RFDR. From Bickel and Rahal (2019).

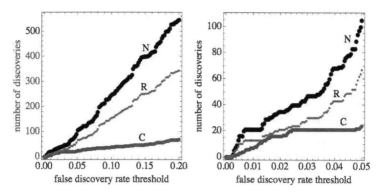

FIGURE 6.2 Number of genes discovered to be differentially expressed as a function of the threshold of the local ("C"; "R") and nonlocal ("N") false discovery rate estimates. The right plot magnifies the lower-left sixteenth of the left plot. From Bickel and Rahal (2019).

estimators seen in Figures 6.1–6.2 and in the exercises (§6.6) corroborates previous findings for this data set (Bickel, 2012b, Fig. 3) and for other gene expression data (Hong et al., 2009).

6.6 EXERCISES (CFDR0–CFDR3)

Exercise CFDR0. Dr. Rate from Section 6.1 argues that since the procedure he suggested is well defined, it needs no correction. (a) What are the advantages of his approach as opposed to correcting it? (b) What are the advantages of correcting it?

Exercise CFDR1. (a) With Basic DE (§4.2.1) as a guide, create the rules of a game in which Nature generates p values of five genes and Lab estimates the NFDRs and the LFDRs of the five genes according to equations (3.5), (6.3), and (6.4). Lab may choose between the estimated NFDR, the CFDR, and the RFDR—after calculating all three—when deciding whether to claim a gene is differentially expressed. (b) Play the game alone or with classmates. (c) Which estimates would have been most successful for Lab?

Exercise CFDR2. Neuhaus et al. (1992) tested 15 null hypotheses pertaining to thrombolytic-treatment outcomes, yielding these

p values: 0.0001, 0.0004, 0.0019, 0.0095, 0.0201, 0.0278, 0.0298, 0.0344, 0.0459, 0.3240, 0.4262, 0.5719, 0.6528, 0.7590, 1.000 (Benjamini and Hochberg, 1995). (a) What achieved FDR estimate, CFDR, and RFDR correspond to each hypothesis? (b) Which hypotheses are very improbable? Which numbers led you to that conclusion? (c) How are the patterns in your estimates based on the treatment data similar to those based on the gene expression data of Figures 6.1–6.2? How are they different?

Exercise CFDR3. Using the Wolfram CDF files of http://davidbickel.com/genomics/ (Bickel, 2019b), (a) find settings of the tuning parameters for which the FDR is misleading, and (b) give the advantages and disadvantages of estimating the LFDR instead of the FDR. (c) Compare and contrast your answers with those of Exercise CFDR0.

CHAPTER 7

The $\mathcal{L}$ Value: An Estimated Local False Discovery Rate to Replace a *p* Value

"Our quest is practically finished. I shall call with the King tomorrow, and with you, if you care to come with us. We will be shown into the sitting-room to wait for the lady, but it is probable that when she comes she may find neither us nor the photograph. It might be a satisfaction to his Majesty to regain it with his own hands."

"And when will you call?"

"At eight in the morning. She will not be up, so that we shall have a clear field. Besides, we must be prompt, for this marriage may mean a complete change in her life and habits. I must wire to the King without delay."

We had reached Baker Street and had stopped at the door. He was searching his pockets for the key when someone passing said:

"Good-night, Mister Sherlock Holmes."

There were several people on the pavement at the time, but the greeting appeared to come from a slim youth in an ulster who had hurried by.

"I've heard that voice before," said Holmes, staring down the dimly lit street. "Now, I wonder who the deuce that could have been."

7.1 WHAT IF I ONLY HAVE ONE p VALUE? AM I DOOMED?

The empirical Bayes methods of hypothesis testing highlighted in previous chapters work best when there are many hypothesis tests, each of which corresponds to a different p value. Fortunately, some of those methods can be adapted to the case of a single hypothesis test.

In fact, the widely applicable lower bound on the posterior probability of the ith null hypothesis out of d null hypotheses from Section 2.3.2 may be seen as a lower bound on the local false discovery rate:

$$\text{LFDR}\left(p\left(x_i\right)\right) \gtrsim 2.7 \times \left|\ln p\left(x_i\right)\right| \times p\left(x_i\right) \qquad (7.1)$$

in the notation of Section 3.3, assuming that the prior probability of the null hypothesis is 50% and that $p(x_i)$ is very small. An important advantage is that it only requires a single p value $(d = 1)$. Calculating $2.7 \times \left|\ln p\left(x_i\right)\right| \times p\left(x_i\right)$ demonstrates how misleading even a small p value can be when interpreted by itself, even when the prior probability of the null hypothesis is low.

However, only knowing that LFDR $\left(p\left(x_i\right)\right)$ is *at least* some number is of limited use in hypothesis testing since we would want to reject the null hypothesis only if we knew it is *less than* some number (recall Exercise L3(c)). The purpose of this chapter is to provide a simple method of estimating LFDR $\left(p\left(x_i\right)\right)$ on the basis of one or more p values $(d \geq 1)$.

7.2 THE $\mathcal{L}$ VALUE TO THE RESCUE!

Given $p(x_i)$, a single two-sided p value, let $z(x_i)$ denote the $\left(1 - p(x_i)/2\right)$-quantile of the standard normal distribution. For example, if $p(x_i) = 0.005$, then $z(x_i) \approx 2.8$, as can be seen from typing one of these commands:

Software	Command that generates $z(x_i)$ from $p(x_i) = 0.005$
Wolfram Alpha (http://bit.ly/2wJOizG)	`-1.41*InverseErfc (2*(1-0.005/2))`
Mathematica (Wolfram Research, Inc.)	`Quantile[Normal Distribution[],1-0.005/2]`
R (R Development Core Team, 2008)	`qnorm(1-0.005/2)`

An approximate $\mathcal{L}$ *value* is then defined by

$$\mathcal{L}\left(p(x_i)\right) = \frac{1}{1 + 1/\mathcal{L}^{\text{odds}}(p(x_i))}, \qquad (7.2)$$

where $\mathcal{L}^{\text{odds}}\left(p(x_i)\right)$ is the approximate $\mathcal{L}$ odds:

$$\mathcal{L}^{\text{odds}}\left(p(x_i)\right) = 1.86\,|z(x_i)|\,e^{-\frac{(z(x_i))^2}{2}}, \qquad (7.3)$$

assuming $|z(x_i)| \geq 1$. Whereas $\mathcal{L}\left(p(x_i)\right)$ estimates LFDR $\left(p(x_i)\right)$ on the basis of $p(x_i)$ alone, $\mathcal{L}^{\text{odds}}\left(p(x_i)\right)$ estimates the prior odds that the null hypothesis is true, as can be seen from solving equation (7.2) for the $\mathcal{L}$ odds:

$$\mathcal{L}^{\text{odds}}\left(p(x_i)\right) = \frac{\mathcal{L}\left(p(x_i)\right)}{1 - \mathcal{L}\left(p(x_i)\right)}. \qquad (7.4)$$

Like formula (7.1), those estimators only work directly when the prior probability that the null hypothesis is true is equal to the prior probability that it is false. However, even in other cases, the

Bayes factor, the likelihood ratio between the null hypothesis and the alternative hypothesis, is estimated by the $\mathcal{L}$ *odds*.

For each prior probability π_0 that the null hypothesis is true, let π_0^{odds} denote the prior odds that the null hypothesis is true:

$$\pi_0^{\text{odds}} = \frac{\pi_0}{1 - \pi_0}.$$

A reasonable estimate of the posterior odds that the null hypothesis is true is the π_0-*prior $\mathcal{L}$ odds*, defined as

$$\mathcal{L}_{\pi_0}^{\text{odds}} \left(p \left(x_i \right) \right) = \mathcal{L}^{\text{odds}} \left(p \left(x_i \right) \right) \pi_0^{\text{odds}}.$$

The π_0-*prior $\mathcal{L}$ value* is

$$\mathcal{L}_{\pi_0} \left(p \left(x_i \right) \right) = \frac{1}{1 + 1/\mathcal{L}_{\pi_0}^{\text{odds}} \left(p(x_i) \right)}, \tag{7.5}$$

which estimates LFDR $\left(p \left(x_i \right) \right)$, as you will show in Exercise LV6. In analogy with equation (7.4), the π_0-prior $\mathcal{L}$ odds is

$$\mathcal{L}_{\pi_0}^{\text{odds}} \left(p \left(x_i \right) \right) = \frac{\mathcal{L}_{\pi_0} \left(p \left(x_i \right) \right)}{1 - \mathcal{L}_{\pi_0} \left(p \left(x_i \right) \right)}, \tag{7.6}$$

estimating the posterior odds that the null hypothesis is true even when $\pi_0 \neq 1/2$.

7.3 THE MULTIPLE-TEST $\mathcal{L}$ VALUE

If there are two or more p values corresponding to two or more hypothesis tests ($d \geq 2$), the first step is the same as before: each $p \left(x_i \right)$ must be transformed to a $z \left(x_i \right)$, as explained in Section 7.2. After that, the *d-test $\mathcal{L}$ odds*, an estimate of the Bayes factor, is approximated as

$$\mathcal{L}^{\text{odds}} \left(p \left(x_i \right); d \right) = \widehat{\sigma}_1 e^{-\frac{\left(1 - 1/\widehat{\sigma}_1^2 \right) \left(z(x_i) \right)^2}{2}},$$

where $\widehat{\sigma}_1$ is known as the *maximum-likelihood estimate of the standard deviation* of all the $z(x_i)$ values:

$$\widehat{\sigma}_1 = \sqrt{\frac{1}{d}\left((z(x_1))^2 + (z(x_2))^2 + \cdots + (z(x_d))^2\right)}.$$

Accordingly, the approximate *d-test* π_0-prior $\mathcal{L}$ odds $\mathcal{L}_{\pi_0}^{\text{odds}}(p(x_i); d)$, an estimate of the posterior odds, and the approximate *d-test* π_0-prior $\mathcal{L}$ value $\mathcal{L}_{\pi_0}(p(x_i))$, an estimate of LFDR $(p(x_i))$, are defined by

$$\mathcal{L}_{\pi_0}^{\text{odds}}(p(x_i); d) = \mathcal{L}^{\text{odds}}(p(x_i); d)\,\pi_0^{\text{odds}},$$

$$\mathcal{L}_{\pi_0}(p(x_i); d) = \frac{1}{1 + 1/\mathcal{L}_{\pi_0}^{\text{odds}}(p(x_i); d)}.$$

The corresponding LFDR estimate given $\pi_0 = 1/2$ is the approximate *d-test* $\mathcal{L}$ value,

$$\mathcal{L}(p(x_i); d) = \mathcal{L}_{1/2}(p(x_i); d) = \frac{1}{1 + 1/\mathcal{L}^{\text{odds}}(p(x_i); d)}.$$

7.4 BIBLIOGRAPHICAL NOTES

Sellke et al. (2001) pointed out that an upper bound on quantities related to the LFDR would be more useful than a lower bound. A $p(x_i)$ threshold of 0.005 is recommended by Johnson (2013) and others. Benjamin et al. (2018) argue that the prior odds that a null hypothesis is true is often 10:1. Equation (7.7) gives a known lower bound of the Bayes factor (Held and Ott, 2016). Bickel (2019c) derived equation (7.3). The Wolfram Alpha page http://bit.ly/2wJOizG was accessed 6 September 2017.

While the LFDR is usually understood as a quantity for multiple testing procedures in the context of hundreds or thousands of p values, its definition as an unknown posterior probability (Efron, 2010) widens its scope. On the basis of a single p value, Bickel (2013) estimated a nonlocal false discovery rate, and Bickel (2014b,

2017, 2019c) proposed various estimators of local false discovery rates. Two simple estimators are explained in Chapter 6.

7.5 EXERCISES (LV1–LV9)

Exercise LV1. Use the $\mathcal{L}$ value or a π_0-prior $\mathcal{L}$ value to revise your answer to Exercise L3(c).

Exercise LV2. (a) Show whether or not the lower bound in formula (7.1) is always less than the $\mathcal{L}$ value when $p(x) \leq 0.3$. *Hint:* Use Wolfram Alpha (http://bit.ly/2wJOizG), a spreadsheet app, or a graphical calculator to plot the lower bound and the $\mathcal{L}$ value together on the y-axis versus the p value on the x-axis. (b) For what values of $p(x)$ is that lower bound *less* than the $\mathcal{L}$ value? For what values of $p(x)$ is that lower bound *greater* than the $\mathcal{L}$ value? (c) Do your answers for part (b) agree with formula (7.1)'s assumption that $p(x_i)$ is very small? Explain why or why not. *Hint:* Not all p values less than 0.3 satisfy the assumption that the p value is *very* small. (d) According to your reasoning for part (c), how small is "very small"?

Exercise LV3. Demonstrate that $\mathcal{L}^{\text{odds}}(p(x_i))$ is approximately equal to $\mathcal{L}(p(x_i))$ when $p(x_i)$ is small enough.

Exercise LV4. (a) Show that $\mathcal{L}_{1/2}(p(x_i)) = \mathcal{L}(p(x_i))$, except for large values of $p(x_i)$. (b) What number is $\mathcal{L}_0(p(x_i))$ equal to? Does that make sense? (c) What number is $\mathcal{L}_1(p(x_i))$ equal to? Does that make sense? (d) Starting with equation (7.5), show that equation (7.6) is true.

Exercise LV5. The calculations needed for this exercise are based on the 15 p values of Exercise CFDR2. (a) Which p values probably correspond to false null hypotheses according to $\mathcal{L}(p(x_i))$? (b) If the prior odds that a null hypothesis is true is 10:1, which p values probably correspond to false null hypotheses according to $\mathcal{L}_{\pi_0}(p(x_i))$? (c) Which is based on more reasonable assumptions, (a) or (b)? Defend your answer.

Exercise LV6. Assuming that $\mathcal{L}\left(p\left(x_i\right)\right)$ is a reasonable estimate of LFDR $\left(p\left(x_i\right)\right)$ when $\pi_0 = 1/2$, demonstrate that $\mathcal{L}_{\pi_0}\left(p\left(x_i\right)\right)$ is a reasonable estimate of LFDR $\left(p\left(x_i\right)\right)$ when $\pi_0 \neq 1/2$. *Hint:* First show that $\mathcal{L}^{\mathrm{odds}}\left(p\left(x_i\right)\right)$ is a reasonable estimate of the Bayes factor.

Exercise LV7. (a) Demonstrate that

$$\mathcal{L}^{\mathrm{odds}}\left(p\left(x_i\right);1\right) = \left|z\left(x_i\right)\right|e^{-\frac{\left(z\left(x_i\right)\right)^2 - 1}{2}} = \sqrt{e}\left|z\left(x_i\right)\right|e^{-\frac{\left(z\left(x_i\right)\right)^2}{2}}$$

$$\approx 1.65\left|z\left(x_i\right)\right|e^{-\frac{\left(z\left(x_i\right)\right)^2}{2}}. \tag{7.7}$$

(b) Using the p values 0.05, 0.01, 0.005, 0.001, and 0.0001, compute $\mathcal{L}\left(p\left(x_i\right);1\right)$, $\mathcal{L}\left(p\left(x_i\right)\right)$, and their difference.

Exercise LV8. (a)–(c) Rework Exercise LV5 using $\mathcal{L}\left(p\left(x_i\right);15\right)$ and $\mathcal{L}_{\pi_0}\left(p\left(x_i\right);15\right)$ instead of $\mathcal{L}\left(p\left(x_i\right)\right)$ and $\mathcal{L}_{\pi_0}\left(p\left(x_i\right)\right)$. (d) Quantify the differences between your results for Exercise LV5(a)–(b) and Exercise LV8(a)–(b). (e) Which results are more reliable? Defend your answer.

Exercise LV9. *DE—Third Edition* has the same rules as Basic DE (§4.2.1) except:

- Nature does not determine observed expression values.

- If the gene is affected (differentially expressed), then Nature records its p value as 1d12%.

- Lab may choose between $\mathcal{L}_{\pi_0}\left(p\left(x_i\right)\right)$ and $\mathcal{L}_{\pi_0}\left(p\left(x_i\right);d\right)$—after calculating both—when estimating LFDR values.

(a) Play DE—Third Edition alone or with others, keeping a record of Lab's $\mathcal{L}_{\pi_0}\left(p\left(x_i\right)\right)$ and $\mathcal{L}_{\pi_0}\left(p\left(x_i\right);d\right)$. (b) Using your knowledge of how Lab generates p values, what would be the best way to estimate the LFDR? (c) For each gene, is $\mathcal{L}_{\pi_0}\left(p\left(x_i\right)\right)$ or $\mathcal{L}_{\pi_0}\left(p\left(x_i\right);d\right)$ closer to the best estimate from the method of (b)? (d) Which estimates would have been most successful for Lab?

Maximum Likelihood and Applications

"When you see a man with whiskers of that cut and the 'Pink 'un' protruding out of his pocket, you can always draw him by a bet," said he. "I daresay that if I had put £100 down in front of him, that man would not have given me such complete information as was drawn from him by the idea that he was doing me on a wager. Well, Watson, we are, I fancy, nearing the end of our quest, and the only point which remains to be determined is whether we should go on to this Mrs. Oakshott to-night, or whether we should reserve it for to-morrow. It is clear from what that surly fellow said that there are others besides ourselves who are anxious about the matter, and I should—"

8.1 NON-BAYESIAN USES OF LIKELIHOOD

8.1.1 Maximum Likelihood Estimation

If $\hat{\theta}$ is the value of θ that maximizes the likelihood function, that is, if

$$L(\theta) < L(\hat{\theta})$$

for any value of θ other than $\hat{\theta}$, then $\hat{\theta}$ is called the *maximum likelihood estimate* (MLE) of θ.

EXAMPLE 8.1
Since the likelihood of Example 2.1 is highest for $\theta = 2$, the MLE is $\hat{\theta} = 2$. However, 2 is not the most probable value of θ, as can be seen from Example 2.3 and Exercise L1.

8.1.2 Likelihood-Based *p* Values

Here, $\hat{\theta}$ is the MLE defined in Section 8.1.1. Let σ denote the standard deviation of $\hat{\theta}$, and let $\hat{\sigma}$ denote the *estimated* standard deviation of $\hat{\theta}$. Then $\hat{\theta}$ is approximately normally distributed with mean θ and variance $\hat{\sigma}^2$; this is abbreviated by writing

$$\hat{\theta} \,\dot{\sim}\, N\left(\theta, \hat{\sigma}^2\right).$$

The abbreviation is the same as in Section 1.2 except that there is a dot above the $\sim$ to indicate approximation rather than exactness. Then the estimate may be standardized by

$$\frac{\hat{\theta} - \theta}{\hat{\sigma}} \,\dot{\sim}\, N(0, 1),$$

that is, the distribution of the standardized estimate is approximately that of the normal distribution with mean 0 and variance 1. It follows that the statistic $\hat{\theta}/\hat{\sigma}$ is distributed as

$$\frac{\hat{\theta}}{\hat{\sigma}} = T \,\dot{\sim}\, N(\phi, 1), \tag{8.1}$$

where

$$\phi = \theta/\sigma.$$

If the outcome of T computed from the data has a value of t, then the p value for testing the null hypothesis that $\phi \leq \phi_0$ is

$$P_{\phi_0}(T > t),$$

which is shorthand for the probability that $T > t$ under the assumption that $\phi = \phi_0$. For example, if $t = 1.8$ and $\phi_0 = 0$, then the p value is

$$P_0 \, (T > 1.8).$$

As with other p values, this p value fails to account for the prior probability of the null hypothesis and thus must be used only with caution (§2.3.1).

If n is sufficiently large, then T^2 has an approximate noncentral chi-square distribution with noncentrality parameter

$$\delta = \phi^2$$

and one degree of freedom. Less technically, knowing the value of the noncentrality parameter would be enough to know $h_\delta \left(t^2\right)$, the density of the test statistic at t^2. Then, using equation (2.1), the likelihood of δ is

$$L\,(\delta) = L_0 h_\delta(t^2), \tag{8.2}$$

and the corresponding likelihood function is L, plotted in Figure 8.1 for features with different values of t^2.

8.1.3 Likelihood-Based Con dence Intervals

Let φ be the random number that has distribution P^t defined by

$$P^t \, (\varphi < \phi_0) = P_{\phi_0} \, (T > t)$$

for all ϕ_0 and t, where T is the statistic given in Section 8.1.2. That P^t is called the *confidence (posterior) distribution* of δ. According to equation (8.1),

$$\varphi \stackrel{\cdot}{\sim} N\,(t, 1).$$

Thus, by Section 1.2,

$$P^t \, (t - 1.96 < \varphi < t + 1.96) \doteq 95\%.$$

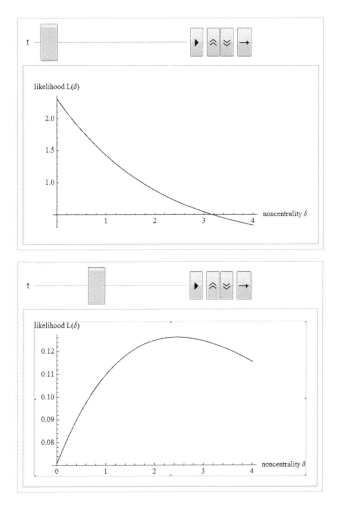

FIGURE 8.1 Likelihood $L(\delta)$ for features with two different values of $t_i = \bar{x}_i/\hat{\sigma}_i$: a lower value in the top plot and a higher value in the bottom plot. The likelihood functions are given by equation (8.2) with equations (4.1) and (8.4).

Defining $\hat{\theta}(x)$ and $\hat{\sigma}(x)$ by the values of $\hat{\theta}$ and $\hat{\sigma}$ computed from the data tuple x and the random θ by

$$\vartheta = \hat{\sigma}(x)\,\varphi,$$

the distribution is

$$P^t \left(\frac{\hat{\theta}(x)}{\hat{\sigma}(x)} - 1.96 < \frac{\vartheta}{\hat{\sigma}(x)} < \frac{\hat{\theta}(x)}{\hat{\sigma}(x)} + 1.96 \right)$$

$$= P^t \left(\hat{\theta}(x) - 1.96\hat{\sigma}(x) < \vartheta < \hat{\theta}(x) + 1.96\hat{\sigma}(x) \right) \doteq 95\%.$$

(8.3)

The set

$$\hat{\theta}(x) \pm 1.96\hat{\sigma}(x)$$

of all possible parameter values between $\hat{\theta}(x) - 1.96\hat{\sigma}(x)$ and $\hat{\theta}(x) + 1.96\hat{\sigma}(x)$ is called an approximate 95% *confidence interval* of θ. Equation (8.3) says there is about a 95% chance that the parameter is in $\hat{\theta}(x) \pm 1.96\hat{\sigma}(x)$.

Notice that $\hat{\theta}(x) \pm 1.96\hat{\sigma}(x)$ is essentially the same confidence interval as the one in equation (4.11) only if there is very little shrinkage, i.e., $B(x) \approx 0$. According to equation (4.7), that only happens when the within-feature variability is much smaller than the between-feature variability.

Thus, the confidence interval $\hat{\theta}(x) \pm 1.96\hat{\sigma}(x)$ is only useful either if there is no plausible null hypothesis such as $\theta = 0$ or if the null hypothesis is probably far from the truth. In other words, if there is a reasonable chance that a null hypothesis is close to the truth, then equation (8.3) should not be used since it has no shrinkage toward the null hypothesis value. Using the terminology of Section 3.2, the non-Bayesian confidence interval $\hat{\theta}(x) \pm 1.96\hat{\sigma}(x)$ can be used effectively if the local false discovery rate is very small but not if it is above 20% or so, in which case the approximate empirical Bayes confidence interval

$$\hat{\mu}_i(x) \pm 1.96 \times \sqrt{1 - B(x)} \times \hat{\sigma}_i(x)$$

would be more reliable, where $\hat{\sigma}_i(x)$ is computed as in Section 8.2.

8.2 EMPIRICAL BAYES USES OF LIKELIHOOD

8.2.1 Error Model for -omics Data

The methods of Section 8.1.2 will now be applied to the problem of determining whether a feature (e.g., gene, protein, or metabolite) is affected by a treatment, disease, or other perturbation. As before, let x_{ij} denote the level of differential abundance of feature i in individual or pair j; thus, x_{ij} is the logarithm of the expression or abundance ratio between the perturbation and the control. Let θ_i be the mean level of differential abundance, and let $\bar{x}_i$ denote the sample mean for feature i, as in Section 4.3.3:

$$\bar{X}_i = \frac{X_{i1} + \cdots + X_{in}}{n}.$$

Further, let $\hat{\sigma}_i$ denote the estimated standard deviation of $\bar{x}_i$:

$$\hat{\sigma}_i^2 = \frac{1}{(n-1)\,n} \left(\left(X_{i1} - \bar{X}_i \right)^2 + \cdots + \left(X_{in} - \bar{X}_i \right)^2 \right). \qquad (8.4)$$

Then a test statistic for the ith feature is

$$T_i^2 = \left(\frac{\bar{X}_i}{\hat{\sigma}_i} \right)^2,$$

which is highly informative about the noncentrality parameter

$$\delta_i = \left(\frac{\mu_i}{\sigma_i} \right)^2.$$

If n is large enough, then T_i^2 has an approximate noncentral chi-square distribution with noncentrality parameter δ_i and one degree of freedom. In other words, knowing the value of the noncentrality parameter would be enough to know the distribution of the test statistic.

The null hypothesis is that feature i is not affected ($\delta_i = 0$). Likewise, the alternative hypothesis is that it is affected, that is, $\delta_i = \delta$

for some nonzero noncentrality parameter ($\delta > 0$). Thus, the value of the noncentrality parameter is

$$
\delta_i = \begin{cases} 0 & \text{if } A_i = 0 \\ \delta & \text{if } A_i = 1. \end{cases} \tag{8.5}
$$

Although the test statistics can be used to get p values for estimating the g_1 probability density function of the p value (§3.2), it is simpler to instead define the relevant probability density functions of $t_1^2, \ldots, t_d^2$, the observed values of the test statistics, i.e., the outcomes of $T_1^2, \ldots, T_d^2$. Let h_0, h_1, and h denote the probability density functions such that the corresponding probability densities satisfy

$$
h\left(t_i^2\right) \approx \frac{P\left(t_i^2 \le T_i^2 \le t_i^2 + \Delta\right)}{\Delta},
$$

$$
h_0\left(t_i^2\right) = h\left(t_i^2 | A_i = 0\right) \approx \frac{P\left(t_i^2 \le T_i^2 \le t_i^2 + \Delta | A_i = 0\right)}{\Delta};
$$

$$
h_1\left(t_i^2\right) = h\left(t_i^2 | A_i = 1\right) \approx \frac{P\left(t_i^2 \le T_i^2 \le t_i^2 + \Delta | A_i = 1\right)}{\Delta}
$$

if Δ is any positive number that is extremely small. As always, the LFDR is the posterior probability that a feature is not affected:

$$
\begin{aligned}
\text{LFDR}\left(t_i^2\right) &= P\left(A_i = 0 | T_i^2 = t_i^2\right) \\
&= \frac{P\left(A_i = 0\right) h\left(t_i^2 | A_i = 0\right)}{h\left(t_i^2\right)}.
\end{aligned}
$$

This LFDR is unknown but can be estimated by the method of maximum likelihood, as will be seen in Section 8.2.3. To do that, the likelihood function must first be obtained.

8.2.2 Likelihood for -omics Data

In order to make the likelihood function as simple as possible, the number of parameters will be reduced to two: π_0, the proportion of features that are not affected, and δ, the value of the noncentrality parameter for the affected features. Recalling that

$$P(A_i = 0) = \pi_0$$

leads to

$$\text{LFDR}\left(t_i^2\right) = \frac{\pi_0 h\left(t_i^2 | A_i = 0\right)}{h\left(t_i^2\right)}, \tag{8.6}$$

which shows how $\text{LFDR}\left(t_i^2\right)$ depends on the first parameter, π_0.

The way in which $h\left(t_i^2\right)$ depends on the second parameter, δ, will now be examined. The law of total probability (Fact 3.1) implies that

$$h\left(t_i^2\right) = P(A_i = 0) h\left(t_i^2 | A_i = 0\right) + P(A_i = 1) h\left(t_i^2 | A_i = 1\right),$$

which, according to equation (8.5), implies that

$$h\left(t_i^2\right) = P(\delta_i = 0) h\left(t_i^2 | \delta_i = 0\right) + P(\delta_i = \delta) h\left(t_i^2 | \delta_i = \delta\right). \tag{8.7}$$

Since the probability density $h\left(t_i^2 | \delta_i = \delta\right)$ is different for different values of δ, equations (8.6) and (8.7) together show how $\text{LFDR}\left(t_i^2\right)$ depends on δ.

To get the likelihood function, equation (8.7) will be written in terms of the parameters π_0 and δ:

$$h_{\pi_0,\delta}\left(t_i^2\right) = h\left(t_i^2\right) = \pi_0 h\left(t_i^2 | \delta_i = 0\right) + (1 - \pi_0) h\left(t_i^2 | \delta_i = \delta\right). \tag{8.8}$$

If the test statistics $T_1^2, \ldots, T_d^2$ are independent of each other, then the probability density at the d-tuple $\langle t_1^2, \ldots, t_d^2 \rangle$ is the product of the d probability densities of all the features:

$$h_{\pi_0,\delta}\left(t_1^2, \ldots, t_d^2\right) = h_{\pi_0,\delta}\left(t_1^2\right) \times \cdots \times h_{\pi_0,\delta}\left(t_d^2\right).$$

Then, using equation (2.1), the likelihood of π_0 and δ is

$$L(\pi_0, \delta) = L_0 h_{\pi_0,\delta}\left(t_1^2, \ldots, t_d^2\right), \tag{8.9}$$

and L is the likelihood function for the error model given by $h_{\pi_0,\delta}$ and the observations that $T_1^2 = t_1^2, \ldots, T_d^2 = t_d^2$.

8.2.3 MLE for -omics Data

Given reliable estimates $\hat{\pi}_0$ and $\hat{\delta}$ of the parameters (π_0 and δ), equations (8.6) and (8.8) can be used to estimate the local false discovery rate as

$$\widehat{\text{LFDR}}\left(t_i^2\right) = \frac{\hat{\pi}_0 h\left(t_i^2|\delta_i = 0\right)}{\hat{\pi}_0 h\left(t_i^2|\delta_i = 0\right) + \left(1 - \hat{\pi}_0\right) h\left(t_i^2|\delta_i = \hat{\delta}\right)}.$$

In the method of maximum likelihood (§8.1.1), the values of the estimates are chosen to ensure that the likelihood given by equation (8.9) reaches its maximum.

That is best understood by studying plots of the likelihood function since the continuous (non-integer) nature of the parameters and data make the tabular representation used with dice in Table 1.3 impossible. Figures 8.2 and 8.3 use

$$\langle t_1, t_2, t_3, t_4, t_5, t_6 \rangle = \langle 0.1, 0.5, 0.7, 1, 2, 3 \rangle$$

as the test statistics for the six genes, proteins, metabolites, or other features.

8.3 BIBLIOGRAPHICAL NOTES

Lehmann (1998, §7.7) covers some topics of Sections 8.1.2 and 8.1.3 in more detail. Allison et al. (2002), Pawitan et al. (2005), Muralidharan (2010), Yang, Aghababazadeh and Bickel (2013a), Yang, Li and Bickel (2013b), and others used maximum likelihood to estimate the LFDR.

Schweder and Hjort (2002) and Singh et al. (2005) provide reviews of confidence distributions with some advances. The term *confidence posterior distribution* is newer (Bickel, 2011a).

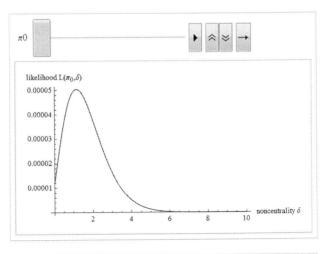

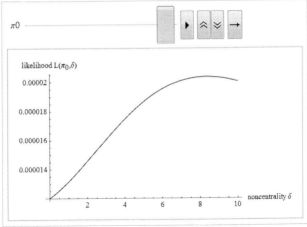

FIGURE 8.2 Likelihood $L(\pi_0, \delta)$ versus δ for two different values of π_0: a low value in the top plot and a high value in the bottom plot. In each plot, some value of δ gives the highest likelihood value. The overall MLEs $\hat{\pi}_0$ and $\hat{\delta}$ can be found by performing such maximization for each value of π_0 between 0 and 1. See Section 8.2.3.

8.4 EXERCISES (M1–M2)

Exercise M1. (a) In general, is the method of maximum likelihood parametric or nonparametric? (b) Is the MLE method of estimating the LFDR parametric or nonparametric? (c) Does the MLE

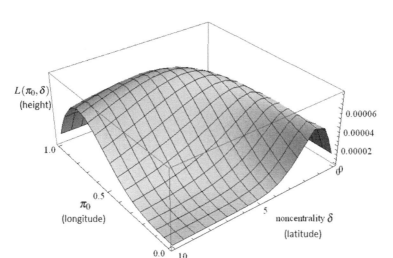

FIGURE 8.3 Likelihood $L(\pi_0, \delta)$ versus π_0 and δ. Here, the likelihood is pictured like the height on a hill with π_0 as the longitude and δ as the latitude. Finding the MLEs $\hat{\pi}_0$ and $\hat{\delta}$ is then equivalent to determining the longitude and latitude of the highest point on the hill. See Section 8.2.3.

method of estimating the LFDR require thousands of features? Explain your answers.

Exercise M2. (a) Suppose that you are a cartographer assigned the task of determining the longitude and latitude of the highest point on a hill shaped like the one pictured in Figure 8.3. What steps would you take to accomplish that using a pen, a notebook, and a portable device that always displays its current height above sea level? (b) What steps would you take to find the MLE of a feature using a pen, a notebook, and a computer program that displays the likelihood $L(\pi_0, \delta)$ every time you type in values for π_0 and δ? (c) How can you use those MLEs of the six features to estimate the LFDR of each of the six features? (d) What can the resulting LFDR estimate for a feature tell you about whether that feature has a different abundance level under the treatment condition than it does under the control condition? (e) What advantages does that LFDR estimate have over a p value for making the determination of differential abundance?

Generalized Bonferroni Correction Derived from Conditional Compatibility

"Good-day, Lord St. Simon," said Holmes, rising and bowing. "Pray take the basket-chair. This is my friend and colleague, Dr. Watson. Draw up a little to the fire, and we will talk this matter over."

"A most painful matter to me, as you can most readily imagine, Mr. Holmes. I have been cut to the quick. I understand that you have already managed several delicate cases of this sort, sir, though I presume that they were hardly from the same class of society."

"No, I am descending."

"I beg pardon."

"My last client of the sort was a king."

"Oh, really! I had no idea. And which king?"

"The King of Scandinavia."

"What! Had he lost his wife?"

"You can understand," said Holmes suavely, "that I extend to the affairs of my other clients the same secrecy which I promise to you in yours."

"Of course! Very right! very right! I'm sure I beg pardon. As to my own case, I am ready to give you any information which may assist you in forming an opinion."

"Thank you. I have already learned all that is in the public prints, nothing more. I presume that I may take it as correct—this article, for example, as to the disappearance of the bride."

Lord St. Simon glanced over it. "Yes, it is correct, as far as it goes."

A.1 A NON-BAYESIAN APPROACH TO TESTING SINGLE AND MULTIPLE HYPOTHESES

The main text of this book has interpreted the p value as a summary of the data for use with Bayes's theorem and the estimation of false discovery rates. This appendix instead presents the p value as a measure of the data's compatibility with the null hypothesis such as the hypothesis that there is no differential expression of a certain gene. To the extent that a p value is low, the tested null hypothesis is incompatible with the data. More positively, to the extent that a p value is high, the tested null hypothesis is compatible with the data. In short, the higher the p value, the more compatible the null hypothesis is with the data.

"Hold on"—interrupts Dr. E. Bayes—"the p value is not so easy to interpret (§2.3.2), for the p value is at best input for calculating a Bayes factor, which is a type of likelihood ratio (§7.2). Even worse, the likelihood ratio cannot be interpreted without a prior probability of the null hypothesis (§2.3.1; Exercise L2). You will just embarrass yourself in genomics data analysis if you ignore the prior probability since it is usually high, sometimes extremely close to 100% (e.g., Exercise L3)."

She has a point. However, the prior probability is not the only way to quantify information possessed before seeing the latest data set. There is also the *prior p value*, which is the *p* value that would measure the compatibility of all previous data sets with the null hypothesis.

Before using that, we need a more inclusive notion of data-hypothesis compatibility, which will in turn require some mathematical notation. Recall that i is a whole number between 1 and d, which is the dimension or the number of genes, SNPs, or other biological features, and that i is used to label a particular feature. For example, $i = 7$ could refer to the seventh gene, more generally the ith feature. Since one null hypothesis is tested for each feature, d is also the number of null hypotheses tested. Before, $p(x_i)$ stood for the p value from testing the ith null hypothesis on the basis of x_i, which records the new measurements of the ith feature. Now the same p value is abbreviated by $p_{\text{null}}(x_i)$ in order to distinguish it from $p_{\text{alternative}}(x_i)$, which is the p value from testing the ith alternative hypothesis on the basis of x_i. For example, if the seventh null hypothesis is that the expression of the seventh gene is unaffected and the seventh alternative hypothesis is that the expression of the seventh gene is affected, then the p values for testing those hypotheses using the gene expression data in x_7 are $p_{\text{null}}(x_7)$ and $p_{\text{alternative}}(x_7)$.

Let y_i stand for all previous measurements relevant to the ith feature, and let $p_{\text{null}}(y_i)$ and $p_{\text{alternative}}(y_i)$ denote the *prior p values*, that is, the p values for testing the null and alternative hypotheses on the basis of y_i. If we combine the previous data with the new data, we get the *posterior p values* $p_{\text{null}}(x_i, y_i)$ and $p_{\text{alternative}}(x_i, y_i)$, for testing the null and alternative hypotheses on the basis of the combined data.

Consider this *either-or condition*: either the ith null hypothesis is true or the ith alternative hypothesis is true. For example, either the expression of the seventh gene is unaffected or it is affected. Now we are ready to quantify compatibility in a way that is not possible with a single p value. The *conditional compatibility* of the ith null hypothesis with the combined data (x_i, y_i) given the either-or condition is

$$
c_{\text{null}}\left(x_i, y_i\right) =
\begin{cases}
\frac{p_{\text{null}}(x_i, y_i)}{p_{\text{alternative}}(x_i, y_i)} & \text{if } p_{\text{null}}\left(x_i, y_i\right) < p_{\text{alternative}}\left(x_i, y_i\right) \\
1 & \text{if } p_{\text{null}}\left(x_i, y_i\right) \geq p_{\text{alternative}}\left(x_i, y_i\right),
\end{cases}
$$

(A.1)

which is called the *conditional c value* for short. If the either-or condition is true, then it is better to reject the ith null hypothesis because $c_{\text{null}}\left(x_i, y_i\right)$ is low than to reject it because $p_{\text{null}}(x_i)$ is low. In other words, instead of rejecting it at significance level α if $p_{\text{null}}(x_i) < \alpha$, do so only if $c_{\text{null}}\left(x_i, y_i\right) < \alpha$. Thus, $c_{\text{null}}\left(x_i, y_i\right)$ may be viewed as a correction of $p_{\text{null}}(x_i)$ based on previous information.

Now for some simplifying assumptions. In light of Dr. E Bayes's remarks, suppose that we have enough experience with the type of genomics data at hand that the previous measurements strongly agree with the null hypothesis but strongly disagree with the alternative hypothesis. In other words, y_i has a very high level of compatibility with the null hypothesis but a very low level of compatibility with the alternative hypothesis. That is represented mathematically by assuming that $p_{\text{null}}(y_i)$ is very high compared to $p_{\text{null}}(x_i)$ but that $p_{\text{alternative}}(y_i)$ is very low compared to $p_{\text{alternative}}(x_i)$. As a result, only x_i has much effect on $p_{\text{null}}(x_i, y_i)$, but only y_i has much effect on $p_{\text{alternative}}(x_i, y_i)$. In short, $p_{\text{null}}(x_i, y_i) \approx p_{\text{null}}(x_i)$ but $p_{\text{alternative}}(x_i, y_i) \approx p_{\text{alternative}}(y_i)$. That simplifies equation (A.1) to

$$
c_{\text{null}}\left(x_i, y_i\right) \approx
\begin{cases}
\frac{p_{\text{null}}(x_i)}{p_{\text{alternative}}(y_i)} & \text{if } p_{\text{null}}(x_i) < p_{\text{alternative}}(y_i) \\
1 & \text{if } p_{\text{null}}(x_i) \geq p_{\text{alternative}}(y_i).
\end{cases}
$$

(A.2)

For practical purposes, that means that as long as the p value from analyzing the new data is lower than the prior p value of the alternative hypothesis, the c value is just the ratio of those two p values. Then the concerns of Dr. E. Bayes are alleviated by choosing a value of $p_{\text{alternative}}(y_i)$ that is low enough to account for the information held before observing the new data. Unless $p_{null}(x_i)$ is even lower, the ith null hypothesis is not rejected. Even if $p_{null}(x_i) < p_{alternative}(y_i)$, the null hypothesis is only rejected at the significance level α if $c_{null}(x_i, y_i) < \alpha$. According to equation (A.2), that is only true if $p_{null}(x_i)$ is less than or approximately equal to $\alpha p_{alternative}(y_i)$.

EXAMPLE A.1

What if, just for fun, we set the prior p value of the alternative hypothesis to be the reciprocal of the number of features? In our notation, that would mean $p_{alternative}(y_i) = 1/d$. Then, according to equation (A.2),

$$c_{null}(x_i, y_i) \approx \begin{cases} dp_{null}(x_i) & \text{if } p_{null}(x_i) < 1/d \\ 1 & \text{if } p_{null}(x_i) \geq 1/d. \end{cases}$$

In that case, $c_{null}(x_i, y_i)$ is approximately equal to the famous Bonferroni-corrected p value, which is $dp_{null}(x_i)$ if $p_{null}(x_i) < 1/d$ and 1 otherwise. That Bonferroni correction for multiple testing was originally designed to control the family-wise error rate (FWER), the probability of rejecting one or more null hypotheses out of a family of d null hypotheses if all of the null hypotheses are true.

In general, the Bonferroni adjustment, like other methods of controlling the FWER, is too stringent to be practical for genomics data analysis (§B.2), as suggested in the caption of Table 3.1. It follows that, in most genomics applications with very large numbers of features, the *generalized Bonferroni correction* represented by equation (A.2) should use a value of $p_{alternative}(y_i)$ that is much greater than $1/d$. In fact, $p_{alternative}(y_i)$ need not depend on d at all.

Dr. E. Bayes laughs, "*Seriously?* Why should I bother with prior p values and conditional compatibility when I can estimate local false discovery rates?"

To be continued … in Appendix B.

A.2 BIBLIOGRAPHICAL NOTES

Bickel and Patriota (2019) studied the general framework of *data-hypothesis compatibility* and abbreviated it by the term *c value*. Their Example 1.1 used a special case of equation (A.1). As Bickel and Patriota (2019) mentioned, its ratio of the *p* values of two hypotheses was proposed by Chuaqui (1991, p. 97).

Posterior *p* values may be obtained from *p* value combination methods such as those of meta-analysis or those of combining confidence distributions (see Singh et al., 2005). Following Edwards (1992)'s concept of *prior support*, Bickel (2012a, §4.3.1) introduced the prior *p* value as a special case of a *subjective p value* (Bickel, 2012a, §4.3).

The connections between compatibility and likelihood run deeper: data-hypothesis compatibility is formally stated under possibility theory (Bickel and Patriota, 2019), which also provides a foundation for novel interpretations of the likelihood function (Dubois et al., 1997; Walley and Moral, 1999; Giang and Shenoy, 2005; Coletti et al., 2009; Maclaren, 2018). That foundation to some extent supports the likelihood paradigm mentioned in Section 2.4 (Bickel, 2014a, 2019d). Bickel and Patriota (2019) cite Mauris et al. (2001, §2.2), Dubois et al. (2004), Masson and Denœux (2006), and Ghasemi Hamed et al. (2012) for interpreting *p* values under possibility theory, and the closely related *s* value is also a possibility measure (Patriota, 2013).

Conditional compatibility (Bickel and Patriota, 2019) uses the version of conditional possibility that is equivalent to the conditional ranking function in Spohn (2012). The reasoning of Spohn (2012) based on ranking functions, coming to life in a Sherlock Holmes example, anticipates the reasoning in Section A.1 based on a prior *p* value and equation (A.2).

Instead of deriving the Bonferroni correction from conditional compatibility, Westfall et al. (1997) interpreted it in terms of Bayes's theorem.

How to Choose a Method of Hypothesis Testing

"Holmes!" I whispered, "what on earth are you doing in this den?"

"As low as you can," he answered; "I have excellent ears. If you would have the great kindness to get rid of that sottish friend of yours I should be exceedingly glad to have a little talk with you."

"I have a cab outside."

"Then pray send him home in it. You may safely trust him, for he appears to be too limp to get into any mischief. I should recommend you also to send a note by the cabman to your wife to say that you have thrown in your lot with me. If you will wait outside, I shall be with you in five minutes."

B.1 GUIDELINES FOR SCIENTISTS PERFORMING STATISTICAL HYPOTHESIS TESTS

Scientists use statistical hypothesis testing not only to learn from data but also to communicate what they learn to some audience of other scientists. Many researchers present testing results in a conference poster or talk and write about those results in grant applications and papers. The successful communicator tailors the message to the audience of peers viewing a poster, listening to a talk, reviewing a paper, or rating a grant.

With that in mind, these simplified recommendations of methods of hypothesis testing are consistent with the reasoning in this book:

1. If you know a prior distribution that would be acceptable to your audience and if you have access to software that can use that prior distribution, then report the posterior probability of each null hypothesis (Chapter 2).

2. Otherwise, if you are testing only one hypothesis, then report a p value with its $\mathcal{L}$ value or another calibration (Chapter 7).

3. Otherwise, if your audience has traditional opinions about correcting p values for multiple testing, then use the Bonferroni method or another method of controlling a family-wise error rate (Example A.1).

4. Otherwise, if you know a prior p value that would be acceptable to your audience, then report the p value and the conditional c value of each hypothesis tested (Appendix A).

5. Otherwise, if your software supports the method of Benjamini and Hochberg (1995) or other false discovery rate methods, then report the achieved false discovery rates with their CFDR or RFDR calibrations (Chapter 6).

6. Otherwise, if you have experience in numerical analysis and have a parametric model that is acceptable to your audience, then write your own code to estimate the local false discovery rates by maximizing the likelihood function (Chapter 8).

7. Otherwise, use previous R code to estimate the local false discovery rate (§3.4).

Algorithm B.1 How to choose a method of hypothesis testing. "BH" is a common abbreviation for the method of Benjamini and Hochberg (1995). See the enumerated list in Appendix B for an elaboration and for cross references to material covered in this book.

```
if(You know a prior distribution acceptable
  to your audience & you have its software.)
    then {Use a fully Bayesian method.}
else if(You are testing only one hypothesis.)
    then {Report a p value with its L value
      or another calibration.}
else if(Your audience has traditional
  opinions about multiple comparison
  procedures.)
    then {Use the Bonferroni method or
      another method of controlling an
      FWER.}
else if{You know a prior p value acceptable
  to your audience.}
    then {Report the p value and the c value
      of each hypothesis tested.}
else if(Your software supports BH or other
  FDR methods.)
    then {Report the achieved FDRs with their
      CFDR or RFDR calibrations.}
else if(You know numerical methods &
  have a model acceptable to your audience.)
    then {Write your own code to estimate an
      LFDR by MLE.}
else {Use previous R code to estimate the
  LFDR.}
```

Those guidelines are summarized in the pseudocode of Algorithm B.1.

The recommendations expose the role scientific audiences play in science. An audience may use its power more responsibly by learning how to evaluate the most common approaches to null hypothesis testing. For example, readers of papers and grants may better evaluate the claims of the authors after transforming p values to $\mathcal{L}$ values (Chapter 7) or c values (Appendix A) and after transforming achieved FDRs to CFDRs or RFDRs (Chapter 6).

Many novel approaches to statistical data analysis that would not convince current scientific audiences are beyond the scope of the guidelines. In particular, the pure likelihood methods cited in Section 2.4 are excluded.

B.2 BIBLIOGRAPHICAL NOTES

The recommendation of Algorithm B.1 regarding a method of controlling a family-wise error rate reflects the consensus that such methods are overly stringent for genomics data analysis. See, for example, Dudoit and van der Laan (2008), who explain how to generalize those methods for the analysis of genomics data. In the viewpoint of Mayo and Cox (2006), methods controlling family-wise error rates or false discovery rates, while useful for decision making, are not suitable for "evaluating specific evidence."

Is a fork better than a knife? That cannot be answered without some context. In the same way, different utensils for multiple hypothesis testing serve different purposes (Bickel, 2011a, §3.1). Advantages and disadvantages of the four major approaches to multiple testing are summarized in Bickel (2013, Table 1).

Bibliography

Aghababazadeh, F. A., Alvo, M., Bickel, D. R., 2018. Estimating the local false discovery rate via a bootstrap solution to the reference class problem. *PLoS ONE*, 13, e0206902.

Allison, D. B., Gadbury, G. L., Heo, M., Fernandez, J. R., Lee, C.K., Prolla, T. A., Weindruch, R., 2002. A mixture model approach for the analysis of microarray gene expression data. *Computational Statistics and Data Analysis*, 38, 1–20.

Benjamin, D. J., Berger, J. O., Johannesson, M., Nosek, B. A., Wagenmakers, E. J., Berk, R., Bollen, K. A., Brembs, B., Brown, L., Camerer, C., Cesarini, D., Chambers, C. D., Clyde, M., Cook, T. D., De Boeck, P., Dienes, Z., Dreber, A., Easwaran, K., Efferson, C., Fehr, E., Fidler, F., Field, A. P., Forster, M., George, E. I., Gonzalez, R., Goodman, S., Green, E., Green, D. P., Greenwald, A. G., Hadfield, J. D., Hedges, L. V., Held, L., Hua Ho, T., Hoijtink, H., Hruschka, D. J., Imai, K., Imbens, G., Ioannidis, J. P. A., Jeon, M., Jones, J. H., Kirchler, M., Laibson, D., List, J., Little, R., Lupia, A., Machery, E., Maxwell, S. E., McCarthy, M., Moore, D. A., Morgan, S. L., Munafó, M., Nakagawa, S., Nyhan, B., Parker, T. H., Pericchi, L., Perugini, M., Rouder, J., Rousseau, J., Savalei, V., Schönbrodt, F. D., Sellke, T., Sinclair, B., Tingley, D., Van Zandt, T., Vazire, S., Watts, D. J., Winship, C., Wolpert, R. L., Xie, Y., Young, C., Zinman, J., Johnson, V. E., 9 2018. Redefine statistical significance. *Nature Human Behaviour*, 2, 6–10.

Benjamini, Y., Hochberg, Y., 1995. Controlling the false discovery rate: A practical and powerful approach to multiple testing. *Journal of the Royal Statistical Society B*, 57, 289–300.

Bickel, D. R., 2004. Error-rate and decision-theoretic methods of multiple testing: Which genes have high objective probabilities of differential expression? *Statistical Applications in Genetics and Molecular Biology*, 3, art. 8.

Bickel, D. R., 2011a. Estimating the null distribution to adjust observed confidence levels for genome-scale screening. *Biometrics*, 67, 363–370.

Bickel, D. R., 2011b. A predictive approach to measuring the strength of statistical evidence for single and multiple comparisons. *Canadian Journal of Statistics*, 39, 610–631.

Bickel, D. R., 2012a. A frequentist framework of inductive reasoning. *Sankhya A*, 74, 141–169.

Bickel, D. R., 2012b. Game-theoretic probability combination with applications to resolving conflicts between statistical methods. *International Journal of Approximate Reasoning*, 53, 880–891.

Bickel, D. R., 2012c. The strength of statistical evidence for composite hypotheses: Inference to the best explanation. *Statistica Sinica*, 22, 1147–1198.

Bickel, D. R., 2013. Simple estimators of false discovery rates given as few as one or two p-values without strong parametric assumptions. *Statistical Applications in Genetics and Molecular Biology*, 12, 529–543.

Bickel, D. R., 2014a. *Model Fusion and Multiple Testing in the Likelihood Paradigm: Shrinkage and Evidence Supporting a Point Null Hypothesis*. Working paper, University of Ottawa, deposited in uO Research. http://hdl.handle.net/10393/31897

Bickel, D. R., 2014b. Small-scale inference: Empirical Bayes and confidence methods for as few as a single comparison. *International Statistical Review*, 82, 457–476.

Bickel, D. R., 2016. Correcting false discovery rates for their bias toward false positives. Working paper, University of Ottawa, deposited in uO Research. http://hdl.handle.net/10393/34277

Bickel, D. R., 2017. Confidence distributions applied to propagating uncertainty to inference based on estimating the local false discovery rate: A fiducial continuum from confidence sets to empirical Bayes set estimates as the number of comparisons increases. *Communications in Statistics – Theory and Methods*, 46, 10788–10799.

Bickel, D. R., 2019a. *A Husband's Translation of Proverbs 31 LXX*. Web page, accessed 16 May 2019. http://dawningrealm.org/2019/05/16/a-husbands-translation-of-proverbs-31-lxx/

Bickel, D. R., 2019b. Interactive comparison of false discovery rates and local false discovery rates. Open access software. https://doi.org/10.5281/zenodo.3234944

Bickel, D. R., 2019c. Sharpen statistical significance: Evidence thresholds and Bayes factors sharpened into Occam's razor. *Stat*, 8 (1), e215.

Bickel, D. R., 2019d. The sufficiency of the evidence, the relevancy of the evidence, and quantifying both with a single number. Working paper. https://doi.org/10.5281/zenodo.2538412

Bickel, D. R., Patriota, A. G., 2019. Self-consistent confidence sets and tests of composite hypotheses applicable to restricted parameters. *Bernoulli*, 25 (1), 47–74.

Bickel, D. R., Rahal, A., 2019. *Correcting false discovery rates for their bias toward false positives.* Communications in Statistics-Simulation and Computation. doi: 10.1080/03610918.2019.1630432.

Blume, J. D., 2011. Likelihood and its evidential framework. In: Bandyopadhyay, P. S., Forster, M. R. (Eds.), *Philosophy of Statistics.* North Holland, Amsterdam, pp. 493–512.

Carlin, B. P., Louis, T. A., 2009. *Bayesian Methods for Data Analysis,* Third Edition. Chapman & Hall/CRC, New York.

Chuaqui, R., 1991. *Truth, Possibility and Probability: New Logical Foundations of Probability and Statistical Inference.* North-Holland Mathematics Studies. Elsevier Science, Oxford.

Coletti, G., Scozzafava, R., Vantaggi, B., 2009. Integrated likelihood in a finitely additive setting. In: *Symbolic and Quantitative Approaches to Reasoning with Uncertainty. Lecture Notes in Computer Science.* Springer, Berlin, Vol. 5590, pp. 554–565.

Doyle, A. C., 1999. The Adventures of Sherlock Holmes. Project Gutenberg, Urbana, Illinois, Accessed 29 May 2019. https://www.gutenberg.org/ebooks/1661

Dubois, D., Foulloy, L., Mauris, G., Prade, H., 2004. Probability-possibility transformations, triangular fuzzy sets, and probabilistic inequalities. *Reliable Computing,* 10 (4), 273–297.

Dubois, D., Moral, S., Prade, H., Jan. 1997. A semantics for possibility theory based on likelihoods. *Journal of Mathematical Analysis and Applications,* 205 (2), 359–380. doi:10.1006/jmaa.1997.5193.

Dudoit, S., van der Laan, M. J., 2008. *Multiple Testing Procedures with Applications to Genomics.* Springer, New York.

Edwards, A. W. F., 1992. *Likelihood.* Johns Hopkins Press, Baltimore.

Efron, B., 2008. Simultaneous inference: When should hypothesis testing problems be combined? *Annals of Applied Statistics* 2, 197–223.

Efron, B., 2010. *Large-Scale Inference: Empirical Bayes Methods for Estimation, Testing, and Prediction.* Cambridge University Press, Cambridge.

Efron, B., Tibshirani, R., 2002. Empirical Bayes methods and false discovery rates for microarrays. *Genetic Epidemiology,* 23, 70–86.

Efron, B., Tibshirani, R., 2007. On testing the significance of sets of genes. *Annals of Applied Statistics*, 1, 107–129.

Efron, B., Tibshirani, R., Storey, J. D., Tusher, V., 2001. Empirical Bayes analysis of a microarray experiment. *Journal of the American Statistical Association*, 96, 1151–1160.

Genovese, C., Wasserman, L., 2003. Bayesian and frequentist multiple testing. In: Bernardo, J. M., Bayarri, M. J., Berger, J. O., Dawid, A. P., Heckerman, D., Smith, A. F. M., West, M. (Eds.), *Bayesian Statistics 7: Proceedings of the Seventh Valencia International Meeting*. Oxford University Press, Oxford, June 2–6, 2002, pp. 145–161.

Ghasemi Hamed, M., Serrurier, M., Durand, N., 2012. Representing uncertainty by possibility distributions encoding confidence bands, tolerance and prediction intervals. In: *Scalable Uncertainty Management: 6th International Conference, SUM 2012, Marburg, Germany, September 17–19, 2012, Proceedings. Lecture Notes in Computer Science*. Springer, Heidelberg, Vol. 7520, pp. 233–246.

Ghosh, J. K., Delampady, M., Samanta, T., 2006. *An Introduction to Bayesian Analysis: Theory and Methods*. Springer, New York.

Giang, P. H., Shenoy, P. P., 2005. Decision making on the sole basis of statistical likelihood. *Artificial Intelligence*, 165, 137–163.

Held, L., Ott, M., 2016. How the maximal evidence of p-values against point null hypotheses depends on sample size. *American Statistician*, 70 (4), 335–341.

Hong, W.J., Tibshirani, R., Chu, G., 2009. Local false discovery rate facilitates comparison of different microarray experiments. *Nucleic Acids Research*, 37 (22), 7483–7497.

Jeffreys, H., 1948. *Theory of Probability*. Oxford University Press, London.

Johnson, V., 2013. Revised standards for statistical evidence. *Proceedings of the National Academy of Sciences of the United States of America*, 110 (48), 19313–19317.

Karimnezhad, A., Bickel, D. R., 2018. Incorporating prior knowledge about genetic variants into the analysis of genetic association data: An empirical Bayes approach. *IEEE/ACM Transactions on Computational Biology and Bioinformatics* (accepted), early access. https://ieeexplore. ieee.org/document/8436435/

Lehmann, E. L., 1998. *Elements of Large-Sample Theory*, 1st Edition. Springer, New York, Corrected third printing in 2004.

Lewis, C. M., 2002. Genetic association studies: Design, analysis and interpretation. *Briefings in Bioinformatics*, 3, 146–153.

Maclaren, O. J., 2018. *Is Profile Likelihood a True Likelihood? An Argument in Favor*. arXiv preprint, arXiv:1801.04369.

Masson, M.H., Denœux, T., 2006. Inferring a possibility distribution from empirical data. *Fuzzy Sets and Systems*, 157 (3), 319–340.

Mauris, G., Lasserre, V., Foulloy, L., 2001. A fuzzy approach for the expression of uncertainty in measurement. *Measurement*, 29 (3), 165–177.

Mayo, D. G., Cox, D. R., 2006. Frequentist statistics as a theory of inductive inference. In: Rojo, J. (Ed.), *Optimality: The Second Erich L. Lehmann Symposium. IMS Lecture Notes – Monograph Series*, Vol. 49, pp. 77–97.

Morris, C. N., 1983. Parametric empirical Bayes inference: Theory and applications. *Journal of the American Statistical Association*, 78, 47–55.

Muralidharan, O., 2010. An empirical Bayes mixture method for effect size and false discovery rate estimation. *Annals of Applied Statistics*, 4, 422–438.

Neuhaus, K. L., von Essen, R., Tebbe, U., Vogt, A., Roth, M., Riess, M., Niederer, W., Forycki, F., Wirtzfeld, A., Maeurer, W., 1992. Improved thrombolysis in acute myocardial infarction with front-loaded administration of alteplase: Results of the rt-PA-APSAC patency study (TAPS). *Journal of the American College of Cardiology*, 19, 885–91.

Newton, M., Quintana, F., den Boon, J., Sengupta, S., Ahlquist, P., 2007. Random-set methods identify distinct aspects of the enrichment signal in gene-set analysis. *Annals of Applied Statistics*, 1, 85–106.

Patriota, A. G., 2013. A classical measure of evidence for general null hypotheses. *Fuzzy Sets and Systems*, 233, 74–88.

Pawitan, Y., Murthy, K., Michiels, S., Ploner, A., 2005. Erratum: Bias in the estimation of false discovery rate in microarray studies (bioinformatics vol. 21(20) (3865–3872)). *Bioinformatics*, 21, 4435.

R Development Core Team, 2008. *R: A Language and Environment for Statistical Computing.* R Foundation for Statistical Computing, Vienna, Austria.

Rahlfs, A., Hanhart, R., 2006. *Septuaginta: Revised Edition.* Deutsche Bibelgesellschaft, Stuttgart.

Royall, R., 1997. *Statistical Evidence: A Likelihood Paradigm.* CRC Press, New York.

Schweder, T., Hjort, N. L., 2002. Confidence and likelihood. *Scandinavian Journal of Statistics*, 29, 309–332.

Sellke, T., Bayarri, M. J., Berger, J. O., 2001. Calibration of p values for testing precise null hypotheses. *American Statistician*, 55, 62–71.

Singh, K., Xie, M., Strawderman, W. E., 2005. Combining information from independent sources through confidence distributions. *Annals of Statistics*, 33, 159–183.

Spohn, W., 2012. *The Laws of Belief: Ranking Theory and Its Philosophical Applications*. Oxford University Press, Oxford.

Strimmer, K., 2008. fdrtool: a versatile R package for estimating local and tail area-based false discovery rates. *Bioinformatics*, 24 (12), 1461–1462.

Strug, L., 2018. The evidential statistical paradigm in genetics. *Genetic Epidemiology*, 42, 590–607.

Strug, L., Hodge, S., Chiang, T., Pal, D., Corey, P., Rohde, C., 2010. A pure likelihood approach to the analysis of genetic association data: An alternative to Bayesian and frequentist analysis. *European Journal of Human Genetics*, 18, 933–941.

Strug, L. J., Hodge, S. E., 2006a. An alternative foundation for the planning and evaluation of linkage analysis. I. Decoupling 'error probabilities' from 'measures of evidence'. *Human Heredity*, 61, 166–188.

Strug, L. J., Hodge, S. E., 2006b. An alternative foundation for the planning and evaluation of linkage analysis. II. Implications for multiple test adjustments. *Human Heredity*, 61, 200–209.

Subramanian, A., Tamayo, P., Mootha, V. K., Mukherjee, S., Ebert, B. L., Gillette, M. A., Paulovich, A., Pomeroy, S. L., Golub, T. R., Lander, E. S., Mesirov, J. P., 2005. Gene set enrichment analysis: A knowledge-based approach for interpreting genome-wide expression profiles. *Proceedings of the National Academy of Sciences of the United States of America*, 102, 15545–15550.

Walley, P., Moral, S., 1999. Upper probabilities based only on the likelihood function. *Journal of the Royal Statistical Society, Series B: Statistical Methodology*, 61, 831–847.

Wellcome Trust Case Control Consortium, 2007. Genome-wide association study of 14,000 cases of seven common diseases and 3,000 shared controls. *Nature*, 447, 661–678.

Westfall, P. H., Johnson, W. O., Utts, J. M., 1997. A Bayesian perspective on the Bonferroni adjustment. *Biometrika*, 84, 419–427.

Yang, Y., Aghababazadeh, F. A., Bickel, D. R., 2013a. Parametric estimation of the local false discovery rate for identifying genetic associations. *IEEE/ACM Transactions on Computational Biology and Bioinformatics*, 10, 98–108.

Yang, Zssss., Li, Z., Bickel, D. R., 2013b. Empirical Bayes estimation of posterior probabilities of enrichment: A comparative study of five estimators of the local false discovery rate. *BMC Bioinformatics*, 14, art. 87.

Index

Printed and bound by CPI Group (UK) Ltd, Croydon, CR0 4YY

22/10/2024

01777838-0002